Culture and Global Change

Linking Levels of Analysis
Emilio F. Moran, Series Editor

Culture and Global Change

Social Perceptions of Deforestation in the
Lacandona Rain Forest in Mexico

Lourdes Arizpe, Fernanda Paz,
and Margarita Velázquez

Ann Arbor
THE UNIVERSITY OF MICHIGAN PRESS

Copyright © by the University of Michigan 1996
All rights reserved
Published in the United States of America by
The University of Michigan Press
Manufactured in the United States of America
⊚ Printed on acid-free paper

1998   1997   1996   3   2   1

A CIP catalog record for this book is available from the British Library.

Library of Congress Cataloging-in-Publication Data

Arizpe, Lourdes.
    [Cultura y cambio global. English]
    Culture and global change : social perceptions of deforestation in
the Lacandona Rain Forest in Mexico / Lourdes Arizpe, Fernanda Paz,
and Margarita Velázquez.
        p.   cm. — (Linking levels of analysis)   **03535695**
    Includes bibliographical references (p.   ).
    ISBN 0-472-10652-X (hardcover : alk. paper). — ISBN 0-472-08348-1
(paperback : alk. paper)
    1. Deforestation—Mexico—Lacandona Forest. 2. Deforestation—
Social aspects—Mexico—Lacandona Forest. 3. Lacandona Forest
(Mexico) I. Paz, Fernanda. II. Velázquez, Margarita. III. Title.
IV. Series.
SD418.3.M6A7513 1996
304.2'8—dc20   95-44333
                                                        CIP

Contents

Now, all of a sudden, they're worried about the forests, but they never used to be. Why don't they knock down the big buildings in the cities so they can plant trees? The government should also worry about its metropolis. Do you think it's right that there are arms factories instead of using the money to improve lives? Nature is the Earth's lungs; we should all help to preserve it, but all of us together, not just us rural folk but also those in the big cities.

María Hernández, farmer's wife, Nuevo Chihuahua

We fall into the necessity of having to cut down trees to feed our families. No matter how much news, propaganda, dialogues, awareness, we get, it is still necessary to cut down trees for food or for money. . . . We've got the information; we understand the problem. . . . The only reality is that we are living in the present, so that our children can have food and schools so that they'll have a future and be more aware.

Eligio Corona, farmer of communal land in Quiringuicharo

The thing is, if the government doesn't bring in order, this is going to be hell. My husband has a communal land lot, and he says that, if the government brings in order, no one's going to be able to produce anything. 'Cause there are lots of spongers who just come into the forest to see how they can get rich. They're the ones who encourage the farmers to organize marches, so they can hide behind them. 'Cause the ones who get arrested and beaten are the farmers and the spongers stay out of sight. They're the ones who sell timber (illegally), and liquor, and they're the ones who want to go on cutting down the forest so they can go on making money, and, if anyone tries to stop them, they start screaming that there is government repression. Actually, it would be a good thing if they were repressed, so they will stop getting the farmers drunk. Luckily, the present governor is very aware of everything. He knows who wants to work the land and who wants to make money by selling timber (illegally) or forest animals to turn all this into a wasteland and then leave, because they have all gone off to live in Tuxtla, in Villahermosa, and they don't care if everything here gets destroyed. The only ones left are the ones who've been deceived, the ones who bought their plots of land thinking they'd be able to live off them, and, when they can't, they turn against the government, not against those who cheated them; these get off scot free. And this is what I say is going to destroy us in the end—the fact that they never admit who's really guilty; they always look for someone else to put the blame on, even when they're the ones to blame. We need consciousness, much more consciousness, but the crooks prefer to blame the government to go on living like rich people. I don't want to sound negative, but that's the way I see things. If people don't wake up, no matter what the government does, nothing will be achieved. You can see it—they're still destroying the forest.

Isaura Arcos, primary schoolteacher, Tenosique

It's all a matter of greed. People aren't happy with what they've got. They all want more, as though they would lose their life if they didn't get a tractor, or a truck, as if they're not worth anything unless they have a Ford. They're insatiable. That's what machismo does. Poor, pardon my language, jerks, since they feel worthless, they'll say I'll show you I've got money, and they'll throw away three million pesos betting on horses. I've seen it. They go to the whorehouses with their wads of money and get robbed out of their minds. . . . If that's not the case, then you tell me where did all the money they got from timber go to? We all went around like millionaires selling timber. . . . No, I got nothing left after the binge, nor do they. That's why they got all worked up, because they don't let them earn money without working. They want to cut down five trees one day to spend the next ten drinking, and there you can see their children all skin and bone, dressed in rags and playing in the dirt, while their dad's in the bar with women of the happy life. Do you think that's right? Maybe I'm getting old, but I can see that's no way to progress.

Aurelio Fernández, *ejido* owner from Nuevo Guerrero

Building a Sustainable Future

This book had just been published in Spanish when the neo-Zapatistas rose up in the region where the research had been carried out. They came out of the mists of the Lacandona rain forest at dawn on 1 January 1994, rifles in hand, and demanded an end to their poverty, to discrimination against their Indian identities, and to the corruption and authoritarianism in the government. These demands are not new, yet the urgency that their senseless deaths gave to them is.

The neo-Zapatistas were clearly acting within the framework of a larger political agenda, yet this book could be considered a "chronicle of a much-announced insurrection," paraphrasing the title of one of Gabriel García Márquez's novels, *Crónica de una muerta anunciada* (*Chronicle of a Death Foretold*), in which everyone knows a murder is about to be committed, yet everyone procrastinates until it happens.

In interviews during fieldwork we were told, literally, about a group training to take up arms. There were areas of the rain forest that we were gently persuaded not to visit, although the actual site of the neo-Zapatista rebellion, the region of Guadalupe Tepeyac and Ocosingo, is on the other side of the Montes Azules Biosphere Reserve from where we were doing fieldwork. The grievances of the farmers in the rain forest are real enough, as they themselves explain in the pages that follow, but other grievances are based more on their perception, in relative terms, of their poverty in comparison with other groups in Mexican society.

Undoubtedly, one of the major issues that sparked off the rebellion were government efforts to stem the destruction of the rain forest land to develop a system of sustainable agriculture. Significantly, the neo-Zapatistas barely mentioned this issue in their communiqués to the media. One interpretation was that they did not want to alienate environmental groups or international public opinion, but their views are mirrored in the quotations here from rain forest farmers.

Given this context, we hope that this book will encourage people in countries of the north—which have been so slow in reducing their greenhouse emissions, thus preserving rain forests as greenhouse gases sink—to understand how detrimental it is for inhabitants of these rain forests to be unable to manage tree felling freely. Many of them say it is a matter of preventing hunger, insecurity, and loss. So much so that they will resort even to insurrection when faced with the prospect of their children dying from malnutrition and curable diseases. It will take more than environmental goodwill to solve the dual problem of over-consumption by some and underconsumption by others, and it will take many negotiations within a global framework.

Toward a Global Framework

As we move into a new world era, major issues are being framed in a new discourse. It would seem that, as humankind passes historical thresholds, new challenges arise that, like a surprise ending in a story, throw a different light on the achievements of world civilization. Never before have democracy and human rights been so broadly embraced, yet never have problems of national and international governance loomed so ominously on the horizon. If we interpret this as the final hegemony of an idea then history has ended, yet neither the idea nor its application seem even close to being universalized; in other words, history may just be beginning. In a similar way, never have such absolute numbers of people enjoyed material well-being, yet never have so many people lived in poverty. The accompanying debate on why the poverty rate has not improved—whether this is due to economics or to population growth—is still inconclusive. And, finally, another threshold shows that the species that gave the planet a sense of history, through what we now know has been a misuse of its riches, risks putting an abrupt end to this history, senselessly.

It is undeniable that progress has been achieved, but mainly it has benefited certain countries, certain elite groups, certain cultures, and one sex. The ethical threshold that we must face at the end of the twentieth century is one in which that progress reaches everyone, which is, in theory, the seminal idea of democracy. Selective progress for some seems to have been intellectually and politically sustained by ego-centered frames of reference, which, for example, considered Europe as the only source of civilization, males as the only agents of change, and the human species as the unique, Promethean protagonist of planetary events. Eurocentrism, androcentrism, and anthropocentrism are slowly being revised and new forms of nation-states, gender relations, and ecosocial interactions are beginning to be built. To give them coherence, a revised frame of reference is needed, one that is grounded on a global perspective to provide a new sense of locality, of place, and of interdependence in a world that is also a planet.

A global perspective would emphasize the *interrelationships* between human beings in different social and economic settings and how such interrelationships, in turn, create patterns of behavior of humans toward the biosphere and the geosphere. Most previous cultures and civilizations were not so out of touch with their natural base of sustenance as industrialism is in its present phase. Nor had the number of human beings and the amount of the Earth's resources that are consumed multiplied as quickly as in the latter part of this century. This complex web of relationships has created a global world whose progress leads inexorably toward planetary geobiochemical changes.

The mystery in all this lies in how each individual's infinitesimal, everyday actions combine to weave an intricate network of global interactions. The wish to penetrate this mystery inspired the research for this book. The question it raises is crucial: To what extent do human beings have the right or power to use and manage the planet's resources? Significantly, it is becoming clear that, in this respect, many traditional cultures have better answers and more explicit rules about interacting with one another and with the planet than modern national legislations. Not even theologians among all the different religions have specific answers; they are only just beginning to return to their Scriptures to seek new interpretations. If there are no divine instructions on the issue of our interrelatedness (and, even if there were, the world's current laicism would make their universal application difficult), it is up to scientists and policy makers to create new laws and institutions to use and manage a world that is being affected by global change.

It is well-known, however, that the uncertainties within scientific research that have been highlighted in recent years have made scientists hesitate to make the first move. And politicians are trapped between expectations of an increasingly high standard of living for a growing population, the vested interests of the powerful, and the in-

creasingly apparent effect of the depletion and erosion of natural resources. One can always look to new technologies to replenish these
resources, but there are no guarantees that this will be achieved as
rapidly and as effectively as needed.

One problem is the lack of cultural, ethical, economic, and political
norms to cope with the world as one world, as the United Nations
Commission on Environment and Development pointed out in its 1987
report. These norms are only now beginning to be created, principle by
principle, word by word, in every transaction and negotiation. A first,
giant step in this process was, of course, taken at the Earth Summit held
in Rio de Janeiro in June 1992, yet the march that follows has been slow.

We hope that this book will contribute to this belabored, intricate
process, for which a myriad of transactions will be needed. It is important that such analysis and debate be conducted with the greatest possible number of participants—scientists and politicians, peasants and
citizens, inhabitants of the North and the South, the East and the West.

The Cultural Dimension of Global Change

As a means of contributing to these transactions, this study analyzes the
cultural dimension of the destruction of the natural environment—in
this case, the deforestation of the tropical rain forest in the Lacandona
area in southeastern Mexico on the border with Guatemala.

Chapter 1 examines the different approaches of the natural and the
social sciences toward globalization and global change. Chapter 2 traces
the history and recent exploitation and colonization of the Lacandona
rain forest. The exploitation of "green gold," as Jan de Vos has called
timber extraction (1988), since the last century was followed by the
recent immigration, both voluntary and government induced, of thousands of landless farmers and Indians into the Lacandona region.

Chapter 3 begins an exploration of the concept of sustainability by
asking whether it is an attribute of culture and analyzing the different
fundamental views that local Lacandona inhabitants have about nature.
We found indifference about and incomprehension toward deforestation, but also anxious concern among both lay people and religious
leaders; they are beginning to create a new social imagery regarding the
nature of the forest and humans' responsibility toward it.

This is further developed in chapters 4 to 6, in which rich interview
data as well as data from a survey carried out in seven communities in

the region are analyzed, with special attention paid to differences in perceptions of local environmental phenomena between mestizos and Indians, men and women, farmers and cattle ranchers, government officials and community leaders. At the same time, their perceptions of the views and intentions of those outside the rain forest, particularly city environmentalists, federal government institutions, the World Bank, and the "international community" are assessed.

Social Perception of Deforestation

In an era of interpretative anthropology and critical theory, "cul-turemes" can no longer be analyzed as though they were pieces of flotsam and jetsam, as they were in classical ethnography. Contemporary research shows that it is essential to know who has emitted the message and why the phenomena under study are being perceived as they are.

We start from a definition of *perception* as "the direct experience of the environment . . . and the indirect information received [by an individual] from other individuals, science, and the mass media" (Whyte 1985:404). We transfer this definition to anthropology, however, to ask what are the cultural and social aspects—what Anne Whyte calls the "subject frameworks in decision-making"—that shape what a person or group perceives and understands. Our assumption in this study is that, when a problem arises, it generates a social process of perception, creation of knowledge, and understanding that is built on the social exchange of information and on the alliances and conflicts with other individuals and social groups. For this reason, the concept of *social perceptions* is used as the main instrument of analysis to explain how different social groups are relating to environmental issues in the Lacandona rain forest.

The results of the seven-community survey, analyzed in chapter 4, indicate who among those living there perceives deforestation as an imminent danger and what changes they have observed in the local natural environment—i.e., in terms of rainfall, floods, winds, and the well-being of wild animals.

Chapter 5 analyzes the assessment by *ejido* farmers, cattle ranchers, housewives, teachers and other professionals, and civil servants of who are the most guilty and vulnerable to the effects of deforestation and who should be put in charge of preventing the destruction of the rain forest. Chapter 6 then presents the different positions taken by each group in relation to the transactions that will have to be carried out to save this "last refuge," as the Lacandona forest has been called (UNAM, 1991).

The book, then, describes an ongoing process whereby social groups take up positions and strategies in a constantly shifting map of social perceptions. It is important to draw up this map and understand its changing patterns for two reasons. First, because socially and politically sustainable solutions to environmental problems can only be successfully negotiated if the differences of perception and assessment of such problems between diverse social and gender groups are carefully understood. Second, because it is necessary to understand the *relative positions* that diverse social and gender groups take in a rapidly changing situation of use and management of scarce resources.

Anthropology, initially created for the study of isolated, stable cultures, now has to develop new analytical concepts and models to study constantly shifting cultures. Since we are "waking a path as we walk" (*haciendo camino al andar*), as the Spanish poet Antonio Machado would say, our findings are an exploration of a research problem in which we are using experimental tools to draw up a map to guide further research and urgently needed policy decisions. The neo-Zapatista uprising attests to the urgency of the situation.

One of the last remaining tropical rain forests in Mexico is being lost; the forest that appears in a Landsat image on the border with Guatemala has already become a deforested strip. And the people living in it, as the interviews show, more than ever are painfully aware of their poverty in comparison with other segments of Mexican society. Hence, the grassroots support in some of these communities for the neo-Zapatistas.

Is there still time for us to achieve conservation with sustainable development for communities in the forest? We believe that a lasting solution must be based not only on capital investments, production, and technology but also on understanding and agreements. Agreements become difficult, however, when economic, political, and ethnic protagonists do not listen to one another or refuse to cooperate in finding trade-offs that are acceptable to all. This research, at the very least, provides an opportunity to consider the points of view of all the groups concerned.

An Interdisciplinary Study

We found it significant that people of the Palenque region and the Lacandona forest were not only willing but eager to speak their minds. Rain forest farmers, both women and men, were particularly keen to share their views, since they have few opportunities to be heard. We are especially grateful to the leaders of the Unión de Ejidos Julio Sabines, the Hernández Dávila and Hernández Canseco family, Pablo Gómez, and many others for those long nightly conversations in the forest. We also thank the members of the Unión de Ejidos Fronteriza Sur for its members' willingness to present and explain their points of view. It would be impossible to mention the names of all those who talked to us, but, as promised, their words appear in this book. Their real names are not given because, as anthropologists, we have a responsibility to respect the privacy and integrity of the individuals we interview.

Such extensive fieldwork would not have been possible without the support of the people and groups working in the forest: the Instituto Nacional Indigenista (INI), particularly Joel Heredia and Concepción López; the PASECOP Program of the now defunct Ministry of Urban Development and Ecology, especially Fernando Maza and Rodolfo Díaz and J. M. Mauricio; and the Ministry of Rural Development of the government of the state of Chiapas, notably Ingeniero López Rivera. Among those taking part in this study were Palenque's enterprising young people, Jaime and Belem, Miguel and Nora, and many more to whom we are indebted, such as the Morales family, especially Moisés Morales, a well-known expert on Mayan culture.

This book is dedicated to the young people who have devoted their efforts to helping Mexico's indigenous and peasant communities and who continue to do so. These are teachers, physicians, agronomists, anthropologists, environmentalists, and artists, among others, who work selflessly in remote corners of forests and mountains all over Mexico. It is they who provide the vitality that makes Mexico's social and political life vibrant and capable of renewal.

It is not surprising that the research for this book came out of the enthusiastic work of a movable group of young people. Baptized as "The Biomass," it included—in addition to the coauthors, anthropologist Fernanda Paz and social psychologist Margarita Velázquez—Luisa del Carmen Cámara, agronomist; Verónica Behn, ecologist; Nina Hinke, biomedic; Miguel Amin, biologist; Lorena Llanes, historian; Marcos B. Eschenburg, electronic engineer; Sandra Behn, psychologist; Benjamín Meyer, philosopher; and Rosa María Martínez Rico, veterinarian. Logistical support for this study was provided by María Isabel Cámara. The group was joined by Patricia Bastin, a Belgian painter, who created the "Educational Posters on Forest Conservation and Development," together with Marisol Fernández, also an artist. Through these sets of twenty posters, which were sent to schools and community centers, some of the main questions about deforestation and global change asked by the farmers in the rain forest were answered.

Advice for research was generously given by demographer Ana María Chávez, philosopher Adriana Yañez, and several other colleagues from the Regional Center of Multidisciplinary Research. This research program also included the project on Women, Health, and Environment directed by Magalí Daltabuit with the assistance of Luz María Vargas and Armando Vargas. Taken as a whole, it was a truly collective research project as well as a multidisciplinary forum of debate and an enriching human experience.

This project would not have been possible without the initial encouragement and generous support provided at all times by the director of the Regional Center of Multidisciplinary Research, Raúl Béjar Navarro, who also helped with comments on the text. Victor Urquidi, Julia Carabias, Mercedes Pedrero, and David Moctezuma provided valuable comments for the revision of the text.

We would like to thank José Sarukhán, rector at the National University of Mexico (UNAM), and Aguilar Sahagun, of the Office of Academic Services, for their support and the funding that made the fieldwork, data processing, and publication of this research possible. We are also indebted to the Werner-Gren Foundation for Anthropological Research for additional funding for this project. Likewise, this work would not have been possible without the administrative support of Maria Teresa Herrera.

This book will be "making its own path as it goes along," and we hope that it will be read by many people, in many places. It will have fulfilled its mission if it helps to weave a global network on which to build a sustainable future for all.

Chapter 1

Human Dimensions of Global Change

For the past two decades phenomena that appeared as isolated processes, such as deforestation, ozone depletion, the extinction of some
animal species, and air pollution, are now being perceived as part of a
broader pattern of global change. The challenge for social scientists is to
study local environmental and social changes that are associated with
global trends. This field of study is being carved out as the human
dimensions of global environmental change (Jacobson and Price 1990).
Calling it "global environmental change" points to the fact that there
may also be research fields on global economic change or global cultural change, although it has been suggested that the term *globalization*
be used mainly to refer to such social processes.

The phenomena of global change are rapidly fostering new research collaboration between natural and social scientists and are certainly becoming some of the major issues for international negotiations
in the years to come. As an emerging field, however, there is still much
to be done to throw light on the scope and implications of global
change. Developing countries should take this unprecedented opportunity to take part in the early stages of a research process that should
lead to more balanced and equitable proposals for global development.

The term *global change* is only just beginning to be used and is thus
still in the phase of having its "human dimensions" mapped out. Basi-
cally, the concept of "global" is used to refer to those phenomena that
affect all the Earth's inhabitants. It is also being used, however, to designate an emerging new perspective of world events and to label a new
stage in human civilization, a new global era. In other words, it is being
used to designate both empirical phenomena and a new theoretical
field, which might be the starting point for creating a new scientific and
political paradigm (Burton and Timmerman 1989). There are even those
who would insist that a field of global science be developed.

Our problem in the world today is that we have to describe, analyze, reflect on, and propose solutions and create new institutions all at the same time, a situation that is only comparable to the formative period of industrialism three centuries ago. Yet today our time is running out. As a result, science is not only expected to analyze the causes and impact of global changes but also to suggest effective proposals to direct them with as much certainty as possible. This means there is a need to change the demarcations between the sciences themselves (i.e., the categories of exact, natural, and social sciences) and between them and their "associates for change," as the International Council of Scientific Unions (ICSU) has called governments, nongovernment organizations, international organizations, and the private sector.[1]

In this context, then, how should global change be defined? If we limit ourselves to the idea that only phenomena affecting the entire human race are global, it would only include those involved in biogeochemical cycles, such as the greenhouse effect or the depletion of the ozone layer. Its study, by definition, would fall within the area of the physical and natural sciences.

Yet almost 80 percent of global environmental changes are caused by human actions, and 100 percent of the solutions to these changes also depend on human actions.[2] Therefore, since most biogeochemical phenomena are anthropogenic, the solutions that have to be found are made up of the decisions and actions of individuals and societies. Environmental global change, then, is inseparable from the human dimensions of global change, although the structure of scientific disciplines is biased toward a heuristic division in these different fields of study. The important point, obviously, is that joint models of the physical, natural, and social sciences need to be developed for the study of global phenomena.

But, just as 20 percent of global phenomena are exclusively dependent on biogeochemical causes and not anthropogenic ones, many areas of sociopolitical change associated with globalization are also exclusively dependent on social, economic, political, and cultural factors. Thus, in addition to the basic human well-being that depends on the planet's well-being, there is another area of human well-being that depends mainly on economic, social, and political development.

The field of global studies, then, is made up of two overlapping circles, one of the natural sciences and the other of the social sciences. While they share a concern for global change and globalization, their specific approaches come from one field or the other. The global perspective emerged from the satellite images that gave us the breathtaking photograph of our blue planet, but it is replayed every day in the instant images we get through telecommunications of events on the other side of the planet.

The dilemma posed by the new globality that we now see in images and know from statistics and the fragmented cosmopolitical scheme of segmentary sciences and of nation-states, tribes and localities, was most accurately captured by the United Nations (UN) Environment and Development Commission, in the first sentence of its 1987 report, *Our Common Future:* "The Earth is one, but the world is not." There are some issues, then, that are now unequivocably global, the major one being that a new threshold has been crossed whereby the human species might become extinct if the bioatmospheric conditions needed to sustain human lives atrofies. Faced with this new idea, humans, only too humanly, tend to resort to old behavior patterns, withdrawing into tight-knit fidelities, whose expression might be political, nationalistic, or ethnic. This disjunction is at the core of what social scientists should be studying and politicians organizing in order to bring human behavior patterns into accord with the new global realities.

The Scientific Challenge

To put it simply, the two main scientific challenges related to global change from our point of view involve creating models to include both the natural and social aspects of global phenomena and explaining the interactions between the local and the global levels. The exact and natural sciences study the scales, rates, and forms of interaction of phenomena such as emissions of greenhouse gases; climatic change; ozone depletion; air, rain, and water pollution; loss of biodiversity; soil erosion; and changes in land use. The social sciences should thus focus on studying the human actions that cause these changes, identifying those who are responsible and those who would be most vulnerable to the impact of these global environmental changes and the new forms of political negotiations and world institutions that must be developed to solve these environmental problems at all levels and locations. These research areas fall within the broad fields of styles of development and of the social perceptions and assessments that different nations or groups establish in analyzing their interests in relation to global environmental change.

To determine which direction to take and to identify priority issues in the social science research of global change, the first challenge involves devising a general scheme, or "wiring diagram," such as the one produced by the commission led by Francis Bretherton for physical and natural sciences to study global change. We attempted this exercise at a 1991 workshop on global change led by Bretherton himself, with the participation of fifty natural, physical, and social scientists.[3] The result, preliminary thus far, is a model that does not represent empirical phenomena but, rather, specifies the information requirements that must be supplied by the different sciences in order to study global change.

The second challenge is to construct models that include biophysical and social perspectives of these phenomena. Several groups of scientists are already working on a project of this sort. Their aim is to create models to establish a typology of the planet's most endangered ecosystems, linking them to human actions that affect them and to the sociopolitical processes that sustain and foster these actions.

The third challenge regarding the human dimensions of global change consists of analyzing the relationship between regional and global phenomena. Two problems arise here. The first is that regional phenomena have a certain autonomy in terms of global dynamics, as shown by studies of historical demography (Turner et al. 1990). At present, there are no models that accurately represent the relationship between these two levels. The second problem is that biogeophysical systems and social systems do not always coincide in space and time. At present, data that are obtained, especially through geographical information systems, are georeferential (i.e., they establish matrices based on geographical parameters), yet the social processes that drive them may in fact come from other geographical regions. For example, deforestation in the Lacandona forest is the result of the economic dynamics of the country's southeastern region but also of the political decisions made by the government in Mexico City, of the consumer habits of urban populations, and of the demographic and land tenure trends in other states. Thus, the types of data and analytical models that need to be constructed to explain this phenomenon fully go beyond the limits of georeferenced data on the Lacandona region.

The scientific challenge in the study of global change also includes other fundamental aspects. Among other things, it has to do with the interpretative frameworks being created to contextualize these studies at a time when scientific paradigms are changing. Natural and exact sciences are now interested in studying nonlinear, irreversible, and unstable phenomena, research that will draw them closer to the social sciences, whose focus has always included these features.

A further challenge, already mentioned in previous pages, involves constructing new theoretical models to study societies' that are experiencing increasingly swift, fluid changes. An additional complication is that such changes must not only be studied but managed as well. This is the sociopolitical challenge in the field of global change.

The Sociopolitical Challenge

The main risk is that at the same time we are heading toward "one world," as proposed by the Brundtland Commission, we are also drifting toward many worlds: some are islands of prosperity and have high protectionist walls; others have oceans of poverty that will become the misery of the new millennium. In this context a global perspective may help focus on the need to achieve a more balanced development for all regions and nations, taking into account cultural and geographical specificities.

In any case, one would have to ensure that the call for a single world would encourage creation of new international agreements and institutions and the gradual development of a global civic culture. What should be avoided is allowing this universal mandate, as the word *universal* was used in the past, to veil the reestablishment of an organization that will favor only the most developed countries.

It would be too long to review the well-known globalization processes associated with current economic and social development. Instead, we will concentrate on the topics presently being suggested as top priorities regarding global environmental change from a social perspective. The International Social Science Council's Program on the Human Dimensions of Global Environmental Change has proposed six key topics in its framework for research in this field:[4]

1. social dimensions of the use of resources: production, reproduction, and consumption
2. perceptions and evaluations of global conditions and environmental changes
3. impact of social, economic, and political structures at local, national, and international levels

4. land use
5. energy production and consumption
6. industrial growth

These topics specifically concern problems associated with the use of natural resources and global change. Some of the main issues involved are:

1. the changes in the use of raw materials and imported agricultural products in developed economies that can change patterns of natural resource use in developing countries
2. the challenges for future industrialization in southern countries because of problems posed by fossil fuel consumption and the emission of greenhouse gases, requiring the transfer to clean technology and investment resources
3. the relationship between population growth, increased consumption of natural resources, food production, and emission of pollutants, particularly, the growing flow of "ecological refugees"
4. the possibilities of raising agricultural productivity, taking into account severe problems of soil erosion, pollution linked to the use of fertilizers and pesticides, changes in regional microclimate, and access to agricultural biotechnologies
5. the protection of tropical rain forests with alternatives of productive development for farming communities in these forests
6. the differential vulnerability of various social and ethnic groups to the impact of global environmental phenomena
7. the ways in which environmental changes are perceived and evaluated by different groups involved in political negotiations to solve socioenvironmental problems and their repercussions on democratization processes
8. the cultural values that establish patterns of well-being and consumer habits that influence economic and social development and the role of the mass media.[5]

The Invisible Sociosphere

The social sciences will have to undergo a transformation to meet the challenge of analyzing and explaining global change. Let us begin by pointing out that the "sociosphere," the social counterpart of the terms *geosphere* and *biosphere,* cannot be seen in a photograph of the planet taken from outer space.[6] As a construct, it exists only in our minds. Consequently, and for a number of other reasons, this has created the impression that transformations of the geosphere and biosphere are purely natural phenomena, whereas, in fact, most are the result of human use of the planet's resources.

Physical and natural sciences have made great strides in their knowledge of how, where, and how fast the natural environment's resources are being used. Social sciences should concentrate on why these resources are being used in specific ways—in other words, for what purpose, and, crucial to our finding solutions for a sustainable future, who is using them. It is essential that we identify the network of interests that often hinder programs and actions aimed at achieving sustainable development.

While this field of research is beginning to be known as global change, Martin Price (1989) points out that it has mainly been defined from geocentric and biocentric points of view. In other words, there is a need to move to a more balanced perspective that will include a human and social dimension.

To do this, it is not enough simply to prepare a list of research topics in relation to global change. There is a need for a new global perspective, and there are already a number of proposals within the social sciences for creating this broader view. Burton and Timmerman (1989:302) believe that a "new paradigm" based on what they call "the development of complex systems" should be built. This would not only mean creating a new theoretical research program but also reconsidering some of the epistemological bases of current social sciences. Since these are intrinsically linked to nineteenth-century philosophical principles, new philosophical and ethical norms that are relevant to the present time must be explored. Gallopín (1989) and researchers at the Bariloche Foundation in Argentina have suggested the concept of "global impoverishment" to include both ecological and economic impoverishment as a central process of global change.[7] What criteria can we use in the social sciences, then, to define global phenomena?

Definition of Global Change in the Social Sciences

Natural scientists have already identified the world's most pressing environmental problems in the *International Geosphere-Biosphere Pro-*

gram: A Study of Global Change.[8] G. K. Menon, former president of the International Council of Scientific Unions, has defined global environmental change as follows:

> Man has modified the environment in the process of living and developing over at least two million years, although for most of this time, man's influence on the environment has been on a small, local scale. It is only during the second half of this century that man has acquired the ability to modify the environment on a global scale and not only in relation to local effects, such as pollution. (1989:60)

From this point of view the criterion for the definition of the global concept is one of *scale*. In addition, as Menon mentions in his description of the main purpose of the International Geosphere-Biosphere Program (IGBP), global change involves "interactive physical, chemical and biological processes regulating the Earth's system, the only environment that supports life, the changes being produced in this system and the way man's actions influence them" (1989:60). Thus, the second criterion for defining this concept is the *interaction* between the different components regarded as its components.

If the main criteria used to define global phenomena in natural sciences are scale and interaction, it would be easy to create a parallel scheme of these phenomena in the social sciences, simply by adding another level of magnitude, that of "global," to the already existing phenomena of economic systems, telecommunications networks, and so on.

Significantly, however, the changes regarded as global, as far as human groups are concerned, are not only associated with changes in the speed, density, and scale of interactions but also with modifications in the structure and complexity of these interactions. As Miller (1989:87) points out, in order to explain world environmental change, it is necessary to examine the direct human actions that produce an effect on it as well as the indirect human actions that set in motion complex sequences of events that also affect the environment. The study of the indirect human actions involved in the dynamic of changing political and economic systems is the reason that the development of specific social theory to explain the global level is necessary. Thus, research into global change should go beyond simply measuring what can be seen; it should tackle the more difficult task of establishing the assumptions and parameters for interpreting this change.

To put it bluntly, since there is no history of global phenomena in human experience, not only do we lack the methods to study them; we also have no basic categories or ideas to help us think about them. Thus, the traditional body of theoretical thought in the social sciences offers very few hypotheses to interpret these phenomena, which helps explain why they are so invisible from the point of view of social science. Only one social science, anthropology, has tried to tackle what is not being conceptualized as globality.

The Anthropological Experience

Despite the fact that anthropology ended up as the study of "others"—that is, of the "peoples without history"—in its basic assumptions it is global in scope and intent. This should be understood in the sense that its field of study has covered all the world's peoples in the present as well as the past and in their interaction with the natural environment.

Information gleaned on this last point has shown, among other things, that this is hardly the first time in history that ecological factors have forced civilizations along a given path of development or decline. One example of this would be the Egyptian civilization, whose appearance has been linked to the desertification of neighboring regions, which drew agricultural populations to the fertile Nile valley and led to the emergence of a highly developed civilization (Manzanilla 1989). There are also many examples of complex states whose collapse was the result of political and social uprisings associated with the depletion of surrounding environmental resources, as may have been the case of the Mayan city of Copán (Abrams and Rue 1988).

A different exercise focusing on attaining a global view of human development was the attempt by many anthropologists to draw up a world map of all cultures. The most important of these was based on comparative data from George Murdock's Human Relations Area Files (1967). There are many similarities between the problems of theory and method put forward in Murdock's project and those currently being examined in relation to the data and models for global environmental change; we shall mention the two most important ones.

The main problem is how to establish the heuristic limits of the units of analysis. In cultural analysis this involves deciding whether a culture should be considered a single unit, despite its internal diversity, or whether its subcultures should be considered as separate entities. This methodological decision may alter cultures' statistical distribution and therefore the generalizations that can be made regarding cultural traits or institutions throughout the world. Thus, these boundaries pose not only technical problems of statistical measurement but also theoretical problems related to the definition of existing relationships between the different cultures. There may be many differentiated cultures belonging to an overarching cultural tradition, as was the case in the Mayan region.

The same kinds of problems arise when the human dimensions of global environmental change are studied. How does one establish a relationship between significant social groups and the georeferenced processes of geoatmospheric models? How can political, administrative, national, state, provincial, and municipal boundaries, for which statistical data exist, be used to study environmental depletion? How can ethnic boundaries be related to environmental phenomena?

Global change, associated both with environmental and human phenomena, essentially constitutes the crossing of a historical threshold to enter a new age in human civilization. This transition is difficult to understand and analyze using the classical instruments of social sciences, so fundamental changes in assumptions and theories must be made. Furthermore, we should look toward the future, both to analyze what is happening and also to contribute to the formulation of a new cosmopolitical scheme, a new ethics, and the new social pacts necessary to ensure the sustainability of a global society.[9]

Chapter 2

The Lacandona Rain Forest: No One's Land, Everyone's Land

Don Esteban Rodríguez, an inhabitant of Marqués de Comillas, in the depth of the Lacandona forest, said, though not without a certain amount of mistrust, that the forest had shifted from being "in the ass of the world" to becoming "the eye of the hurricane," a reference to the current, and unexpected, growing concern, both nationally and internationally, over the conservation of this immense, precious tropical forest.

Don Esteban's comment is accurate. Whereas the Lacandona rain forest was regarded as a "no-man's-land" or the "Lacandón desert" until thirty years ago, nowadays, because of the potential global consequences of its destruction, it has become "every man's land."

Some of the factors, related to deforestation, that have alarmed the international community are the disappearance of sources to regulate atmospheric gases, the loss of biodiversity, and the climatic effects caused by the deforestation. All of these will also have devastating consequences for the nearly 200,000 settlers presently living in the rain forest. If deforestation continues, their fertile plots of land will become barren deserts incapable of being revitalized for seed planting, despite the promises of countless fertilizers sold by multinational companies. In fact, one of the main problems already affecting the world's tropical forests is soil degradation caused by limitless tree felling, the abuse of chemical fertilizers and pesticides, and the inappropriate use of land for cultivation.

Several studies mention deforestation as the main cause of the problem, since deforested lands lose their productive capacity in a few years. Although the soil is rich, deforestation hastens the elimination of the fertile topsoil, and the removal of biomass decreases the contribution of organic material, thus hastening erosion (Meave del Castillo 1990:24). According to the UN Development Programme (1990:25–33), annual rates of deforestation were 0.54 percent annually for South America, 1.60 percent for Central America, including Mexico, and 0.4 percent for the Caribbean. The annual rate of deforestation in Mexico for the same period was 1.3 percent annually, which is equivalent to 3,690 square miles per year.

Yet deforestation is not a recent phenomenon. It has occurred for thousands of years all over the planet and for a long time was considered the ultimate goal of human control over nature. Today, however, because of its magnitude, instead of being a source of pride, deforestation has become a cause for alarm. It has been estimated that Central America has lost 38 percent of its forest area, while Africa has recorded losses of 24 percent just during the second half of this century (Schmink 1992:2). Data for Mexico provided by the National Forestry Inventory (Inventario Nacional Forestal de Gran Visión) shows that from 1961 to 1985 there were 28.5 million acres of high and medium forests in the whole of Mexico, whereas by 1991 this figure had fallen to 21.75 million acres, an approximate loss of 24 percent since 1985 (Secretaría de Agricultura y Recursos Hidráulicos [SARH] 1991:32).

Occupying just 6 percent of the Earth's total surface, tropical forests are home to nearly 50 percent of all animal species. Costa Rica, for example, boasts more bird species than the United States and Canada together, while Madagascar has more than two thousand tree species compared to the four hundred species recorded for North America's coniferous forests (Sloan et al. 1988:29). Many of these species are now in danger of extinction.

At the same time, the problem cannot be reduced to purely biological issues, even though these are crucial. Because tropical rain forests are on the whole more susceptible to changes caused by human action and less able to recover its productive capacity, their partial or total degradation would directly affect nearly one billion people currently living in tropical countries and would indirectly affect all the inhabitants of the planet.

In the case of the Lacandona forest a recent study by Wilerson (in Aridjis 1990) shows that in the past thirty years the forest has lost 70 percent of its original area—estimated at 3.5 million acres—and that 18.3 percent of the remaining 30 percent has already been depleted.

Auctioning the Forest

The Lacandona forest's recent history began in the late nineteenth century, with the large-scale exploitation of precious woods. Spanish, Belgian, English, and U.S. companies extracted such woods for the international markets. As in the case of other export-oriented products (coffee, rubber, and bananas), the boom in the markets for mahogany, cedar, and other precious woods was part of the world system whereby Mexico and other former colonial regions became suppliers of raw materials for industrialized nations.

The rapid expansion of nineteenth-century capitalism in Europe and the United States provided the capital needed for new investments with higher profit rates and a greater capacity to absorb exports. In this economic world system Mexico, like other countries in Latin America, Asia, and Africa, first suffered colonial plunder and then exploitation through unequal terms of trade (Mandel 1978:247–87; Ayala and Blanco 1985:17–18). In the nineteenth century in Mexico the adjustment to external demands was accompanied by political and economic shifts at the national and local levels. The Liberal Party implemented policies to transform the country's economic structure so as to permit capital accumulation (Gutelman 1980:48; Thompson and Poo 1985:103–4). Accordingly, the Liberal government tried to free the land and the work force from corporate hold by the church and by Indian communities.

Liberal policies radically altered the country's agrarian structure and led to the proletarization of a vast number of peasants. From 1824 to 1859 legislation was established, affecting first church property and later communal lands to favor the emergence of a middle-class sector that would become the mainstay of the country's modernization (Favre 1984:68–69; Thompson and Poo 1985:100–101). The main beneficiaries of such policies, however, were actually large estate owners, because these laws were used to extend estate properties (Pedrero 1983:33).

At the turn of the century, under Porfirio Díaz's government, the tendency toward concentrating land was further strengthened by the laws of colonization issued from 1883 onward. The state ordered national and foreign settlers to report virgin lands and set up surveying companies. As Gutelman notes (1980:33), "the companies or settlers received a third of the land surveyed by way of payment and enjoyed the option to buy with preferential rates for purchasing the remaining two thirds from the State."

By 1826 Chiapas had already issued the first laws of colonization, which led to the concentration of land at the expense of indigenous communities. Yet the large landed estates created at this time were essentially no different from the old estates; since their production was not aimed at internal or external markets, there was no capital investment, and production relationships continued to be based on servitude. Not until 1875, with the triumph of the Liberals over the Conservatives in this southern state, was the reporting, buying, and selling of uncultivated land given further impetus, and taking possession of land led to the creation of new estates. García de León (1985:173–74) points out that,

> between 1875 and 1908, 27 percent of the total area of Chiapas was reported by private companies. Over 2 million acres of supposedly uncultivated land, 4,532,500 acres, were granted to wood, oil, rubber, and coffee companies, most of them through the English surveying company Mexican Land and Colonization Company (MLCC), represented in Mexico by Luis Huller. . . . Thus from 1806 to 1905, Huller's company surveyed and sold 1,425,840 acres to private owners in Chiapas.

Hitherto uninhabited areas such as the Soconusco and the Lacandona forest became a center of attraction for foreign investors. While in Socunusco most estates became coffee plantations, in the Lacandona region the extraction of precious woods became the main economic activity.[1]

Vast stretches of this land were gradually occupied by timber companies representing the interests of industrialized countries (González Pacheco 1983:54–56). These firms took advantage of the "auctioning off" of the rain forest allowed by the agrarian laws of 1889, 1890, and particularly 1893. These laws, on the one hand, removed the limits to property expansion into uncultivated land and, on the other, canceled owners' obligations to populate and delimit their land (Gutelman 1980:35). As mentioned earlier, the Survey Law for Uncultivated Lands (Ley de Deslindes de Tierras Ociosas) allowed a third of the surveyed land to be obtained free and the rest to be bought very cheaply. After 1902 the acquisition of rain forest lands became even simpler, since the government granted concessions for the exploitation of land in return for a small rent in kind (Gutelman 1980:36; González Pacheco 1983:83).

The timber companies occupying the forest not only used the agrarian laws to set up their plants; they also benefited from the influx of foreign capital. This explains the spectacular increase in exports during this period, as noted by Ayala and Blanco (1985:17): "Between 1877–1878 and 1910–11, the total value of exports rose from 32.5 million to 281.1 million, in other words, an increase of almost 846 percent." According to González Pacheco's analysis (1983:136), during the first year (1877–78) precious woods accounted for 12.5 percent of all products exported and 23.4 percent if one includes logwood (9.5 percent) and mulberry trees (1.4 percent).

Until 1914 the principal market for precious woods was Europe, mainly England. Demand had risen as a result of the expansion of colonial empires and the depletion of European forests. After World War I the United States strategically placed its capital in Latin America and gained control of its raw materials exports. This shift in world capitalist forces had great repercussions for the Mexican economy, as fluctuations in the demand for products and changes in international capital flows directly affected the accumulation model based on exports (García de León 1985:156–58; Velasco 1985:46).

In the case of wood such fluctuations sometimes led to collapse, as, for example, in the case of the Bulnes Hermanos Company. During its thirty-six years in the forest this company exported the most wood to Europe. In 1914 it sent 21,477 cubic yards of wood to England yet was forced to close the following year as a result of the withdrawal of English capital (González Pacheco 1983:101).

After 1914 companies operating in the rain forest were mainly financed by U.S. capital. Perhaps the most representative of these was Romano Company Sucs. From 1880 to 1925, when it ceased operations, it had produced approximately 1,205,100 cubic yards of precious woods (González Pacheco 1983:113). U.S. companies continued to expand in Latin America after World War I, displacing mainly French and British companies in the region.

In Mexico the effects of this international economic restructuring led to a fall in exports and a reduction of credits that affected the economy. Yet during World War II Mexican industry expanded, as domestic demand rose and imports fell (García de León 1985:155–61; Velasco 1985:45–46; Halperin 1986:356–60). Forestry was to supply the raw materials needed for industry. Article 6 of the Forestry Law, issued in 1942, stipulated, "The construction of industrial forestry units to supply raw materials for the mining, paper, construction, transport and arms industry is considered to be in the public's interest" (Ovando 1981:43; González Pacheco 1983:155–56).

During this period forestry entered a new phase in that it was incorporated into the industrialization process. In 1951 the Maderera Maya Company was established as a subsidiary of the U.S. Vancouver Plywood Company and set about buying forest land, including Lacandona rain forest land, through third parties. It purchased a total of 1,093,335 acres of the best land to set up a plywood factory. The federal government made the forestry permit contingent, however, on the establishment of a pulp and cellulose plant. Despite fulfilling all the requirements established by the laws in force, the company never obtained the concession. For fourteen years Maderera Maya attempted to carry out its plans, and, failing to do so, it auctioned off its permits to the Weiss Fricker Mahogany Company, selling its land into private ownership in the late 1960s (González Pacheco 1983:156–79).

Weiss Fricker devised a short-term plan for intensive timber extraction; the company's contracts were only for ten years. Thus, in 1964 it founded Aserradora Bonampak, S.A., which operated until 1973. The following year Weiss Fricker installations were sold to Nacional Financiera, S.A.; this was the beginning of government control of rain forest exploitation.

Government Management of the Rain Forest

In 1972, under President Luis Echeverría, the federal government issued a decree granting 1,535,802.5 acres of the central part of the region to sixty-six Lacandona heads of family in an attempt to cope with increasing encroachment on forest lands. It also established the Compañía Forestal de la Lacandona, S.A. (COFOLASA), as a subsidiary of Nacional Financiera in 1974. The government also created the Montes Azules Biosphere Reserve to stem commercial exploitation of the area and established

the entire upper catchment area of the Tulijá and Usumacinta rivers, on the Mexican side, as a forest reserve for limited exploitation, under the supervision of the Technical Department of the

Subsecretaría Forestal y de la Fauna, with COFOLASA as the only firm entitled to exploit precious wood. (Muench 1982:97)

COFOLASA subsequently negotiated timber extraction with the legal representatives of the Lacandona community, officially recognized as sole owners of the forest. Since, however, other Indian groups, mainly Chol and Tzeltal, had already settled in the region, it was not long before agrarian conflicts arose between these groups and the Lacandona communities. Settlers who had migrated to the forest in the 1950s demanded that the government recognize their rights to the land. One of the pressure tactics used was to prevent the timber company from cutting trees in areas occupied by these settlers. There were several years of conflict, as a result of which several villages were razed to the ground. Then, in 1976, the government proposed including Chols and Tzeltals within the Lacandona community, which led to the establishment of two new villages, Frontera Echeverría, also known as Corozal, and Velasco Suárez, or Palestina.

COFOLASA's performance was not particularly successful. In addition to agrarian problems, it had to cope with high operating costs; extracting the precious wood required road building, as the wood by now could only be found in the most inaccessible areas. In 1980, after years of operating at a loss, the company was declared bankrupt. It was taken over by the government of the state of Chiapas, which ran it until 1989, when it was eventually closed. The firm's history was marked by inefficiency, corruption, and collapse, and it is regarded as a significant factor in the Lacandona forest's current drama of destruction.

Colonization of the Forest

While the establishment of timber and rubber extraction companies in the rain forest during the latter half of the nineteenth and the early twentieth centuries encouraged a number of workers' families to settle in the woodland, it does not explain how the whole region began to be populated. Massive colonization of the Lacandona forest began in the early 1950s and increased between 1964 and 1970 (Lobato 1979; González Pacheco 1983).

The Tzeltals from the municipality of Ocosingo were the first to go deep into the forest in search of land. After founding the village of Lacandón in the municipality of Palenque, they negotiated its legal ownership, which was granted in 1954. Years later, in 1961, Tzeltals from Bachajón achieved the same for the *ejido,* the agrarian reform unit, of Santo Domingo (Lobato 1979:132). After that numerous indigenous groups from the municipalities of Bachajón, Yajalón, Ocosingo, Pantelhó, Tila, Tumbalá, and Salto de Agua, pushed by insufficient lands and demographic pressure, left their homes in search of lands in the rain forest.

Indians from the highly populated highlands of Chiapas were not the only ones to regard the rain forest as an alternative to the shortage of land in their places of origin. In the early 1960s, the Departamento de Asuntos Agrarios y Colonización (DAAC) published a decree in the *Diario Oficial de la Federación.* It nullified all concessions of Mexican land by expropriating; for reasons of public interest, 1,475,410 acres of the Lacandona rain forest had been granted to private owners in order to establish ejidos, bringing landless peasants from the states of Sonora, Coahuila, Chihuahua, Aguascalientes, and Zacatecas (Lobato 1979:81–82).

Based on the 1946 Law of Colonization, the DAAC created seventeen ejidos in the Lacandona area under collective ownership as established in the Agrarian Reform laws. The project was curtailed, however, by two factors. Firstly, the lands allocated by the DAAC to peasants from the northern states were also being requested by Indian migrants who had come to the rain forest in previous years. In view of the conflict that was beginning to brew the government of Chiapas settled the matter in favor of the Tzeltal and Chol Indians. Of the seventeen districts that had been planned, only four were eventually created: Nueva Esperanza, El Lacandón, Once de Julio, and Ricardo Flores Magón.

The state government explicitly said that it was responding to a policy of supporting the poorest indigenous sector. Yet a review of the state's agrarian structure in the early 1960s shows that 2.4 percent of the privately owned production units of more than 2,500 acres occupied 58.2 percent of all farming land, whereas plots with less than 25 acres, accounting for 42 percent of the production units, took up only 0.9 percent of this land (García de León 1985:226–28). We must assume, then, that the settlement in favor of the Indian petitioners mitigated tensions between poor peasants and landowners while maintaining the large estates intact.

The decision by Samuel León Brindis, then state governor, by no means represented a conflict between local and federal power, since in 1962, the year after the DAAC drew up its farming community plan, the 1946 Agrarian Law was repealed. At the same time, Article 58 was added to the 1942 Agrarian Code. It specified that land that had previously belonged to the nation, to states, or to municipalities would be used to form or expand ejidos or to create new ejido population centers (NEPC) (Reyes Osorio et al. 1974:701). Thus, private colonization was prohibited, and León Brindis's decision became legally endorsed.

The fact that the creation of ejido settlements in the Lacandona area coincided with changes in the law was neither a matter of chance nor a purely local phenomenon. The decade of the 1960s, particularly the presidential terms of López Mateos and Díaz Ordaz, was characterized by a shift in agrarian policy. Unlike President Alemán, who had given priority to private agricultural production, modernizing vast stretches of land in northern Mexico and distributing more than ten thousand certificates to guarantee immunity from expropriation, López Mateos supported ejidos, land distribution, and small landowners, since pressure from this peasant sector was already beginning to make itself felt.

New developments in the international scene also encouraged this shift in agrarian policy. Faced with the challenge of the Cuban Revolution, the Alianza para el Progreso (ALPRO) proposed a policy of restructuring agrarian structures in Latin American countries to stem revolutionary movements. Thus, millions of dollars in aid and new, redistributive policies for rural lands were included in López Mateos's Program for Integral Agrarian Reform (Appendini and Salles 1983:164). This policy involved the colonization and distribution of virgin lands (la frontera agricola) in the humid Mexican tropics.

At that time people had no notion of the ecological consequences that might result from massive colonization or the implementation of development projects that were ill suited to tropical lands.[2] Colonizing the tropical rain forest was seen as a means of killing two birds with one stone: (1) it would provide land for landless peasants, thus alleviating agrarian pressures, and (2) it would encourage the production of basic foodstuffs for the domestic market. Obviously, the latter implied that the state would need to invest in infrastructure, which it did not do in the Lacandona forest.

From approximately 1960 to 1974 the northern and southwestern regions of the forest received an influx of hundreds of Indian migrants in an attempt by the government during the 1970s to bring migrants together in large settlements in which minimal services could be provided. In some cases they received a portion of national land, nothing more, as in the case of the migrants in the region of Las Margaritas and its hinterland, where the neo-Zapatistas rose up in early 1994.

Looking back on the history of the forest's colonization, it seems clear that the agrarian authorities used this great tropical forest, with little planning, to relieve agrarian political pressures, especially in central and northern Mexico, which witnessed a "battle for the land" (*la batalla por el campo*), as it has been called, in the 1970s. During the 1960s it was mainly Indian peasants from Chiapas who migrated to the rain forest, and this was considered politically appropriate by the state; in the 1970s, however, rain forest lands were offered to petitioners from other states.

Landless farmers from Guerrero, Puebla, Oaxaca, Michoacán, and Chiapas itself began to settle along the banks of the Lacantún River, in the Marqués de Comillas region, the most remote corner on the border with Guatemala. Except for a group of migrants from Oaxaca and Veracruz, most of them had never had any contact with the tropical rain forest, and their farming experience had been in temperate regions and in some cases on areas with mechanized farming, especially in northern Mexico. "We came here with the idea of farming," said an ejido farmer from Pico de Oro, "sowing corn, beans, and chili and riding on a tractor—that's what I know about."

For these migrants the forest was an unknown, savage jungle that had to be domesticated. Despite the difficulties of adaptation, there were several advantages to making the move: 125 acres of land were handed out to each ejido farmer. Imagine what this meant to a farmer from central Mexico whose whole family eked a living out of a meager 1 or 2 acres of land. Additionally, loans to buy cattle were easily offered by the government banks, plus the government itself, through COFOLASA, would guarantee to buy whatever timber was felled. Such an offer was, indeed, irresistible.

Yet not all the migrants took equal advantage of the offer. While there was a certain degree of homogeneity concerning the migrants' needs, expectations, occupations, and income level in the 1960s, this was not true in the next two decades. Since colonization of the Lacandona rain forest was fostered so indiscriminately, it attracted people of extremely diverse interests. Some were freeloaders only interested in making money from the timber business, and who, in fact, succeeded in doing so by totally clearing their plots of land and later selling their ejido rights. Others were peasants who saw cattle raising as an alternative, given the instantly available credit facilities, and invested time and money in clearing the forest. Still others arrived with nothing more than the hope of a better life and now eke out a living on only 4 or 5 acres of seasonal crops, since the forestry ban issued in 1989 and restrictions on cattle raising credit have curtailed their expectations.

Thus, the forest was up for grabs to relieve agrarian and demographic pressures on Mexico's lands. At the same time, it also led to indiscriminate and occasionally violent plundering by clandestine timber merchants, cattle ranchers, and government officials who were, sometimes, one and the same person or else closely allied partners. Such government officials were supposedly elected by more than 90 percent of the votes cast for the Partido Revolucionario Institucional (PRI), which, it was widely known, were the result of massive fraud. This is the reason one of the main demands of the neo-Zapatistas was for clean and fair elections. In the 1980s this complex interplay of forces was compounded by a further consideration related to national security.

Prevailing conditions in Guatemala in the early 1980s led to incursions into Mexico both by the guerrillas and the Guatemalan army as well as hundreds of thousands of refugees fleeing from the Guatemalan government's scorched-earth policy in Indian regions. In view of the gravity of the situation sectors of the Mexican government debated possible solutions. These ranged from total militarization of the border to the creation of settlements and ejidos that would form a "national human border," so to speak. The latter measure was adopted, and from 1983 onward the colonization of the southern and eastern regions of Marqués de Comillas was further promoted and implemented. Migrants were attracted, as mentioned, through land grants, the opening of dirt roads, and support for agriculture—including free mechanical saws—and cattle raising. Throughout this period ecological considerations that had been fully expressed, studied, documented, and recognized by the Mexican government were overridden by this political concern. As a result, it is estimated that between 1983 and 1987, 197,500 of the 495,000 acres in Marqués de Comillas were deforested.

As in other forest areas, the granting of land in Marqués de Comillas was improvised and badly planned. Not until 1987, in view of the growing conflicts and the renewed concern regarding the environmental destruction, was the Comisión Intersecretarial de la Selva Lacandona formed as a coordinating body to implement policies and programs for a rational use and preservation of the rain forest. That same year the Uniones de Ejidos and the Asociaciones Rurales de Interés Colectivo, that is, farmers organizations operating in the Lacandona rain forest, signed an agreement with the federal and state governments,

> promising to respect the ban (prohibiting tree felling); not to touch the Biosphere Reserve; to use live trees for fencing; to rationalize land use . . . ; and not to fell or clear vegetation in high or medium forest; to stabilize slash-and-burn practices; to develop productive activities compatible with the resource preservation, aimed as much at local self-sufficiency as at capitalizing the ejido; and to assume joint responsibility with the government in the prevention and control of forest fires, tree felling, and illegal trafficking of flora and fauna. (Argueta and Embriz 1990).

At the same time, the government promised to provide advisory, technical and financial support for a regional development that would preserve its natural resources. Two programs were begun to stop deforestation. One was coordinated by the Secretaría de Agricultura y Recursos Hidráulicos (SARH) to plant new precious wood and fruit trees. The other, sponsored by the World Bank, was channeled through the state government; it involved the reforestation of 25,000 acres of forest with rubber plantations, which would also provide needed jobs and income for farming families.

Neither of these two projects has been as successful as the government hoped, with both the farmers and the government officials blaming each other for its shortcomings. The government has pointed out that "the population in Marqués (de Comillas) is not interested in the reforestation programs promoted by the Ministry of Agriculture," as one official complained. Another added, "They want us to take the young trees, whether wood or fruit trees, to their plots of land." The farmers have responded by saying, for example, "What the government wants to do is to reforest, but how many trees am I supposed to carry on my back for 3 or 4 miles?" They also complain that rubber plantations require more financial support than they can give. As we shall see, this pattern of blaming other groups for failures or errors is a common feature in the Lacandona area and is no doubt one of the factors that has limited the possibility of finding a solution acceptable to all groups concerned.

Finally, in 1989, the newly elected government of Carlos Salinas de Gortari decreed a total forestry ban that put an abrupt halt to tree felling throughout the Lacandona area. Governor Patrocinio González Garrido upheld the ban, despite massive opposition, closing down sawmills and enforcing controls in the use of mechanical saws to stop deforestation. These measures increased political tensions in the Lacandona area and even more so in Marqués de Comillas, where incoming migrants along the border and, among them, some fortune hunters had not had time to settle in.

In short, the recent history of the Lacandona forest can be described as anarchic, conflictive, and exploitative. The most diverse interests—political, environmental, and economic—have converged on this geographical area, leading to its current problems. The exact size of the total population inhabiting Mexico's largest and most varied tropical forest is unknown, although it is estimated at approximately 200,000 (González-Ponciano 1990:73). The fact is that these people already live there, and, though they may once have been free to colonize in any way they chose, they are now responsible for its future and its environmental consequences. The fate of the forests now concerns the whole world, and a solution must be found to negotiate a sustainable future for the Lacandona rain forest.

Chapter 3

Culture and Sustainability: Concepts of Nature in the Lacandona Rain Forest

Sustainability is like democracy: a complex and elusive concept but essential to establish a goal horizon that will galvanize human will. Like democracy, it is a means of approaching the world that has to be based on practice. Just as the theory of ethics derives from the social practice of morality, sustainability will gradually acquire an empirical content susceptible to analysis and definition, as the practices to achieve a balance between humanity and nature advance and diversify.

For the time being, we can only begin to formulate basic questions about sustainability. We begin this chapter by comparing the possible meaning of the concept of sustainability from an anthropological point of view with other concepts used in the social sciences. We then relate this term to the concept of nature among the inhabitants of the Lacandona forest in southeast Mexico.

Sustainability and Social Reproduction

Based on a worldwide survey, the United Nations Commission on Environment and Development has defined sustainability as the possibility of "satisfying the needs of the present without compromising the ability of future generations to satisfy their own needs" (1987:43). According to Robert Ayres (1991), the problem with this definition is that no one knows how to measure well-being in social terms. Therefore, he proposes a more complex definition:

> Sustainability is a process of change in which the exploitation of resources, the direction of investments, the orientation of technological development and institutional change are in harmony with and enhance current and future potential in order to satisfy man's needs and aspirations.

Although this definition is much more precise, we are still left with the problem of defining the phrase "in harmony with." This points to the need to involve social scientists, anthropologists, sociologists, and psychologists in exploring this concept.

We stress this because economists have already gone some way toward producing a more operative definition of sustainability. Karl-Goran Maler, for example, states that "economic development in a specific area (region, nation, or world) is sustainable if the total reserve of resources, human, capital, reproducible physical capital, environmental, and non-renewable, do not decrease over time" (1990:240). The Comisión Económica para América Latina (CEPAL) has argued, instead, that the biological concept regarding an activity as sustainable if it does not infringe on certain natural laws should be broadened to include criteria on the management and use of natural resources, such as civic and institutional participation and political decision making (1991:22). It is clear, then, that sustainability should not be defined on the basis of the available amount of natural resources but, rather, on the use that is made of these resources.

The use of natural resources has a great deal to do with the way in which societies are reproduced. In this respect, it is interesting that the *concept of sustainability* should have arisen a few years after the *concept of social reproduction* was discussed in the social sciences, especially since, although they have similar signifiers, they mean two quite different things. The latter is derived from the interest of neo-Marxism in knowing how certain ideological or cultural structures are reproduced through the state or the dominant class, as in the work of Louis Althusser and Antonio Gramsci, or authoritarian cultures are reproduced through symbolic violence in education, as in the work of Pierre Bourdieu and Jean-Claude Passeron. The concept of social reproduction was used to discover why undesirable situations of domination or inequality remain unchanged. This may be useful in studying sustainability, since it becomes necessary to ask why, if humans are beginning to perceive—with uncertainty yet, on the whole, unequivocally—that their actions are sawing the limb on which they are standing, they continue to reproduce behaviors that will deepen this process.

Put another way, methodological individualism leads research to assume, in this case, that environmental destruction is entirely dependent on the myriad of individual actions affecting the natural environmental. Instead, the assumption we make is that sustainability is not only an attribute of the sum of individual behaviors but also of the synergies created by economic and political systems. A theoretical tool such as that of social reproduction could be useful in this respect. The question to ask is, how can styles of development be implemented to ensure that societies reproduce themselves in a sustainable manner?

Such a concept should not be confused with biological reproduction, which is an area usually associated with women. In fact, feminist theory distinguishes between biological reproduction, the replacement of labor, and social reproduction. This distinction may be useful to show that sustainability must be conceptualized both as a problem of relationships between human beings living together in societies and the natural environment as well as relationships within these societies, ensuring a social reproduction that will not increase pressures on the natural environment. In other words, "natural" sustainability must be premissed on "social" sustainability.

If we agree that the internal dynamics of human societies shape the behavior of individuals in those societies toward nature, then we must begin by asking what kinds of concepts those societies have about nature. This was the question that guided the data collection and analysis that we present in this chapter.

Is Sustainability a Cultural Axiom?

Culturally, sustainability means continuity, a wish to prevail and, of course, ethnographies have shown that all societies share the desire to prevail as an identity group even after the deaths of its individual members. No historical or anthropological evidence has been found of any society that has consciously and deliberately walked to its death. In fact, the collective suicide of a religious sect, involving almost seven hundred people in Georgetown, Guyana, in the 1970s, seems to have been a signal of alarm for human civilization.

In view of this, it would be unconscionably shallow to think that sustainability is just one more term in the political discourse at the end of the millennium. Instead, it is associated to the deepest structures of human consciousness and could thus be postulated as an axiom of all cultures. But what happens when cumulative changes in technology, demography, and economy cause a society to live in contradiction with the ecosystem that supports it?

This has happened, as we know, in specific cases throughout history. What we do not know is how the cultural axiom of collective survival was overridden by contrary interests. Nor do we know, in the case of empires or societies that have disappeared, why they were unable to carry out the necessary social, political, and economic transformations quickly enough to be able to restore the balance between the *sociosystem* and the *ecosystem*.

While these questions go beyond the scope of this exploratory study, the question our fieldwork data can help elucidate along these lines is, how are the values of a society toward nature expressed culturally?

Nature, World, Land: **Many Names**

"Nature? I wouldn't know what it is. I'm not educated," said a woman from an ejido in the Lacandona rain forest who now lives in the town of Palenque. A strange reply: for her the word *nature* belongs to the world of schools, from which she and the peasants and indigenous peoples, with their age-old cultures, are certainly excluded. In fact, fieldwork showed that, when the inhabitants of the Lacandona region spoke about the natural environment, they used different terms not only according to their educational level but also as a feature of their ethnic and religious upbringing. Most people, especially peasants, whether mestizo or Indian, referred to the natural environment through the biblical image of "divine creation" and interchangeably used the words *Earth, world, all,* or *things* to refer to it, rather than the word *nature*. A first distinction to be made is between beliefs based on religious or nonreligious concepts of the origin of the world.

The Deistic Version

The deistic version of divine creation can be summed up in the phrase, "God made everything," in answer to the question, "What is the world?" (Que es el mundo?). "There is a creator of all natural things," stated an ejido owner from the Lacantún River. "I don't think it could just have been a matter of chance; a god created the heavens, the earth, the water, and the animals." God's creation of nature is an unquestionable fact for those with religious beliefs. "I don't really believe in evolution," said a peasant from the El Naranjo ejido. "We believe that it was God who made everything. We don't believe in the transformation of times—that's what we were brought up to believe since we were little."

The Bible is a constant point of reference on this subject, and, in the words of David Ortiz, a Catholic Tzeltal ejido farmer from San Miguel: "Well, as we know, like they told us, like the word of God says, everything belongs to God. We look to the Bible, really." One surprising feature, however, is the free interpretation of Genesis expressed by both Catholics and Protestants, Indians and mestizos, to explain the creation of the universe. Bernabé Alvarez, a Chol ejido farmer from San José Babilonia, provides a beautiful, idiosyncratic interpretation of the relevant biblical passages:

> In the beginning, God made the heaven and the Earth; the Earth was covered in water, like an abyss, there was no dry land. The spirit of God flew over the waters like a dove and then went up to heaven and thought, "It's not good like that." That's what God thought, so he made dry land and put a river in the earth, and that's how trees and animals were born and then man, Adam. And now there are fruit trees, too. As man got tired, God took out one of his ribs to make his companion, and then there were two. . . . Then he made fish in the water, when the water was still clean. He made all the animals on dry land, and, when he had finished, God said, "Now it's ready."

The only differences in these accounts of the origin of the world lie in the poetic liberties taken by the narrators. "I'm Protestant," asserted Mateo Alvarez, a Chol peasant.

> I really like the Bible because it says everything. It says that in the olden days there was a huge expanse with nothing in it, and then it was divided up bit by bit into nature, and then man came and then woman. God made everything, and he called it mountain and animals, and that's how it stayed.

Yet, while there is a unity of criteria about the origin of nature among those with religious beliefs, the same does not apply to their destiny. The only response to questions about who created the world and why it was created is to be found in Genesis, but when people refer to *why* it was created and *how* humans should treat the Earth, their answers are not only based on religious criteria but also on cultural, economic, and political ones. This explains the differences among the deists, who can be classified into three main groups: (1) those who think that nature was created to be used, regardless of whether it is destroyed, (2) those who justify its destruction as a result of financial need, (3) and those who explicitly favor conservation.

Let us examine each of these positions more closely. A review of these accounts reveals a surprising feature. Except for those who justify the destruction of the natural environment out of financial need, in the other two groups we find an ambivalent God supporting destruction as much as conservation.

A milkman from the San Mateo ejido pointed out that "we shouldn't kill wild animals because God didn't make them for us to eat; he made them for us to admire. . . . We shouldn't cut down trees because they belong to God; we can only eat the fruit."

In contrast with this position, a Chol ejido farmer from San José Babilonia also justifies his relationship with nature with reference to God: "When God made everything, He said, 'Who's going to be the one in charge?' Then He said to man; 'You're going to be the one in charge. Everything I made is yours, and no one is going to take it away from you.'"

This position was taken up by a day worker, another Chol from the same ejido. He argued that "God made man to work, calm things down, and rule the world. God made nature for us to rule it. It's alright to cut the trees down."

In general, most Chol farmers' accounts adhere to the belief that God created the world for humans to "inhabit, enjoy, work and rule," with no reference whatsoever being made to the conservation of the natural environment. Furthermore, those we interviewed did not seem to regard the environmental depletion as a problem, with one significant footnote: they all repeatedly expressed their concern that "*there isn't enough land anymore*," in the sense that they will no longer have a means of producing food, and so the age-old threat of famine hangs in the air. They also frequently regret the fact that they will have no land to pass on to their children. This is a common concern among peasants and Indians, yet only a few see the connection between this fear and the problem of having too many children. This is because, given their deistic beliefs, they are unable to conceive of not having children as the result of an individual choice. Choice is not part of their cultural heritage, as it has not been part of their political or cultural existence in Chiapas.

Of the rest of the interviewees few referred to the need to preserve natural resources or had any explicit views on the subject. Significantly, those who did were mainly Protestant. For example, "Nature has created God, and he even created us," said a young twenty-year-old Protestant woman, a policeman's wife from Palenque. "Jehovah made nature for us to take care of it and protect it and to give us food. We're a family; like nature, we're united," she noted, adding disapprovingly that "there are people who are just interested in money and that's why they destroy everything. The thing is, they're going to make a lot of money in the beginning, but after a while they won't make anything."

If things are destroyed, it is as a result of human actions, not by God's design. "God created everything—the plants and man—but nature was out there for man. It was made for him to reproduce it, but now man is destroying it. He's already destroyed a lot," said Don Nicolás Díaz, of Chol descent, who now works as a bricklayer in Palenque.

Besides blaming humans, some interviewees spoke of our responsibility to the natural world. One of these was Roberto Gallegos, a farmer from El Naranjo: "We have the obligation and right to take care of nature. Others say that God made nature for us to use, but, if we destroy it, then it won't belong to God anymore."

An interesting feature of these explicitly conservationist accounts within deistic views is the reasons behind them. Some cite aesthetics: "We need trees. They look nice, and they give shade," remarked the owner of a pharmacy in Palenque. As a city dweller, she regards nature as a garden, a totally different perception from that of the peasants whose survival depends on nature. "Our lives depend on nature," said a Tzeltal ejido owner from San Miguel. "If it ever gets destroyed and the land ends up all dried up, we won't have anything left." Thus, most peasants justify conservation in terms of survival.

Finally, it is worth mentioning a Chol peasant's characteristic view that gives an entirely different picture of what is going on:

The Devil took everything there was on Earth and said it was his, and he doesn't like people to grab or kill the animals because they are his. . . . The Devil thinks he owns the world, but God gave us all the animals for us to eat, for food. If we kill lots of animals, God won't get cross, but the Devil will.

This account is interesting because it indicates a free interpretation that runs counter to the references we know and expect of Christian beliefs about the world. It also reminds us that the dangers of nature, particularly the forest, such as death from snakebites or poisonous animals or other dangers, also form part of this natural world and may be identified with demonic forces.

Nonreligious Versions

There is a sharp difference between those with a religious view of nature and those who do not profess any religion. The majority of the latter belong to the urban sector and are civil servants, freelance workers, professionals, office workers, or students. They refer to the natural environment as *nature* rather than *world, Earth,* or *everything.*

"The creation of the world cannot be explained in religious terms," said a sociologist living in Palenque. "It took a long time to make, and an evolutionary process was involved." A civil servant from the same town noted: "We are part of the evolutionary process. We carry out a series of human actions, including birth and death, but, unfortunately, we have no contact with nature." It is further stressed that nature is necessary to human lives: "We live off nature"; "It helps us"; "It lets us lead healthier lives."

Many interviewees lamented that humans beings are destroying nature "out of a lack of awareness." A civil servant put it bluntly: "The problem is man's stupidity," and he continued: "Lots of people don't think, and, if we don't think, society is headed for destruction. The solution is to establish control over our relationship with nature in order to preserve and link awareness with reason." A sign maker in Palenque states it more fully:

> People are not really very aware of how the forest's destruction affects them. They live in their own little world, attacking their own species. Man is not aware of the effects of this destruction or of what he has destroyed. If man has affected nature, its not because he controls it but because he is not aware of what he's doing.

These comments are interesting because, although environmental destruction is perceived as something that will compromise the future of society or of the human species, it is attributed solely to individuals'

lack of awareness; their actions are not placed in the context of economic, political, or social processes.

Yet not everyone having a nonreligious concept of nature shares a
conservationist position nor perceives the risks implied by the total
destruction of the forest. Rubén Merino, an ejido farmer from Nuevo
Orizaba who emigrated from the northern state of Sonora, lives deep in
the Lacandona area. He asked:

> Well, what use is the forest to us? It's just there. It might as well be
> of some use. Otherwise, its just like looking at a calendar. I want to
> work the forest so I can leave my children something, and, if that
> means cutting down the forest, that's what I'll do.

Beliefs and Sustainable Behavior

Of course, beliefs may have little to do with actual behavior toward the
natural environment. In fact, it has recently been argued that culture
has no influence on the processes that are destroying the natural environment. For example, the beliefs of Hinduism strongly support the
natural environment, yet they have been unable to prevent severe ecological damage in India (Rockwell 1992). This argument points to the
fact, as have others, that economic necessity may override any benev-
olent cultural predisposition toward nature.

Although our research was not designed to analyze the actual
behavior of informants toward the forest, fieldwork observation indicated, on the one hand, that there seems to be few differences between
the deforestation behavior by those who express religious views,
whether Catholic or Protestant, and those with nonreligious views on
nature. This would tend to support the hypothesis that economic interests are more powerful than religious or lay beliefs.

A visible difference can be seen, however, between Indians and
mestizos. Indian settlements have preserved trees and plants in the
center of the villages and have grown them in their home compounds
much more than mestizos have. They make more use of the forest's
medicinal plants and other natural fruits, although they hunt or sell
animals in the same way as the mestizos.

The mestizos, especially migrants from highly urbanized areas
with mechanized agriculture have cleared all the vegetation and set out
to urbanize their villages as quickly as possible. They also try to obtain

the greatest possible output, from wood extraction or agriculture, whereas Indian families tend to produce only what they need for their domestic consumption.

On the other hand, both our observations and interviews indicated a much higher degree of passivity of Catholics toward what is happening in the rain forest, simply because they believe that everything is "in the hands of God." But this is also related to the fatalism that has been reported in Indian cultures. Fieldwork observation tended to show that Protestants, and especially those who hold rationalistic views, are much more active in seeking solutions to the destruction of the rain forest.

Chapter 4

The Culture of the Ax, the Machete, and the Sling

The previous chapter presented the Lacandona forest inhabitants' concepts of nature. This chapter describes the ethnography of their perceptions on recent changes in the natural environment, particularly the deforestation, of the Lacandona rain forest.

Two different methodological operations will be involved. Ethnography takes us back to the classic method of anthropology, in which the norms obtained from informants were recorded within a well-defined semantic field. We used this method in fieldwork fully aware of its inherent trend toward "the nostalgia for the origin, the archaic and natural innocence, the purity of the presence and the presence itself of the word," as Derrida very poetically put it (1978, 95). We followed this course, admitting, following Lévi-Strauss, that its lack of acceptable historical significance is offset by the value justifying it as an instrument of method.

We found, however, that norms relating to deforestation did not exist but were, rather, in the process of being created. In a sense, then, there was no normative center in people's utterances. Rather, their views ran along the borders of emerging lines of discourse.

Therefore, our second methodological operation was to map out the conceptual territory of the emerging perceptions of environmental change in the region. The first task was to find out what boundaries were established in our informants' discourse related to such change.

External Pressures, Internal Perceptions

Environmental change, particularly, deforestation, soil erosion, water pollution, and the loss of rain forest flora and fauna have been going on in the Lacandona rain forest for decades, yet until recently they had not been converted into an issue of public life.

Local inhabitants undoubtedly perceived these changes in the sense of "receiving the impression of an object through the senses."[1] Likewise, some of them no doubt understood these phenomena, in the sense of "discovering the substance, quality and relation between things through the exercise of the intellectual faculties."[2] Such perceptions and understanding acquired a social and political significance, however, which turned them into public issues because of information and pressures coming from outside the region.

From the 1960s onward, as we have seen, the government, development agencies, and state entrepreneurs all agreed that the forest was "unproductive," a word still used by many in the region, especially large-scale cattle ranchers. They insist that productivity can only be increased through agriculture and cattle raising, which means doing away with the rain forest.

At the beginning of the 1970s a few environmental groups began to emerge in favor of conservation of the rain forest. They were lead by Gertrudis Duby and Manuel Alvarez del Toro, and their efforts were successful in creating a broad coalition of environmental groups in Chiapas linked to other national and international nongovernmental organizations.

By the end of the 1980s the information conveyed from the outside world to Chiapas by the mass media, the shift in government policies toward the rain forest, the appearance of international agencies interested in conservation, and the multiplication of environmental groups all pressed for conservation. Consequently, deforestation, together with other environmental issues, came to the fore in public discussions in the whole of Chiapas.

To find out how local people were giving semantic form to this change in their perceptions, we conducted a survey of seven communities: two ejido communities settled in the 1970s (Pico de Oro and Reforma); two ejido communities settled in the 1980s (La Victoria and Nuevo Chihuahua), the four located in Marqués de Comillas; two rural communities settled in the 1960s near Palenque (Lacandón and La Unión); and Palenque itself, with specific focus on three sub-communities: one low-income urban group; one high-income urban group; and one cattle-ranching group (more information is provided in the appendix).

"What Do You Think Is the World's Greatest Danger?"

In the answers to the open-ended question "What do you think is the world's greatest danger?" we found at one end of the scale those who consider life itself as dangerous, so that everything is a threat. At the other end were the optimists, who declared that "everything is fine; there's no problem."[3] Between these two extremes there were seven main types of concerns, regarding: war, 23.8 percent;[4] poverty, 13.7 percent; pollution, 8.3 percent; attitudes, 8.3 percent; illness, 7.9 percent; deforestation, 6.7 percent; environmental degradation, 4.4 percent; divine punishment, 4.4 percent, overpopulation, 3.2 percent; and others, 3.5 percent.[5]

One response bears mentioning, which, although unique in the survey, was sometimes mentioned by other informants during fieldwork. This is the view, given by a farmer from Pico de Oro, that the forest itself is dangerous: "Here, we're surrounded by dangers, animals, snakes, falling sticks, and crocodiles." This is, of course, true and worth pointing out to offset the sometimes idyllic and inoffensive image of the rain forest created by the mass media. The inhabitants of the forest have to fear these dangers, and for them conserving the forest could mean preserving these dangers. By way of example, in 1991, the twelve-year-old eldest son of a former ejido commissioner from Pico de Oro died from a snakebite. Experiences such as this necessarily inform the concepts local people have of the forest and, incidentally, of the number of children they wish to have.

Significantly, only 6.7 percent of those surveyed regarded deforestation as the most imminent danger, and even this could have been a response that they thought we wanted to hear.

The differences in the responses to our question were analyzed by region, community, occupation, sex/gender, ethnic group, and religion. Table 4.1 shows the differences by community, with the highest-income groups and cattle ranchers from Palenque giving similar answers, related mainly with pollution and, to a lesser extent, with environmental degradation, overpopulation, and attitudes.

In contrast, the low-income groups from Palenque and the farming communities near Palenque mostly mentioned war, poverty, and attitudes as the greatest dangers facing the world. There is greater variation among rain forest communities, which point to war, deforestation, illness, and divine punishment as pressing dangers.

TABLE 4.1. Percentage Distribution of Answers on the Greatest Danger in the World Today, by Community

Answers	Total	Community					
		PR (1)	VC (2)	LU (3)	PL (4)	PH (5)	CR (6)
Poverty	13.7	15.6	9.4	15.6	18.8	6.3	16.7
Deforestation	6.7	17.7	8.3	—	4.2	2.1	2.1
War	23.8	17.7	17.7	35.4	31.3	31.3	10.4
Pollution	8.3	1.0	2.1	5.2	4.2	25.0	29.2
Illness	7.9	4.2	8.3	13.5	14.6	—	4.2
Environmental degradation	4.4	3.1	6.3	2.1	4.2	8.3	4.2
Overpopulation	3.2	4.2	1.0	—	—	8.3	10.4
Divine punishment	4.4	2.1	11.5	4.2	2.1	—	2.1
Other*	3.5	6.3	5.2	2.1	—	—	4.2
Don't know	15.7	20.8	25.0	14.6	10.4	6.3	4.2
Total	100.0	100.0	100.0	100.0	100.0	100.0	100.0

Key: (1) Pico de Oro and Reforma Agraria, (2) La Victoria and Nuevo Chihuahua, (3) Lacandón and La Unión, (4) Palenque low income, (5) Palenque high income, (6) Palenque cattle raisers
*Includes lack of technology, natural disasters, the Mexican government's ban on tree felling, and natural dangers of the forest.

The contrast also holds if the first four priorities for the answers are compared (see table 4.2). They clearly tend to vary according to dyads: high-income residents and cattle ranchers in Palenque; low-income residents in Palenque, and Lacandón–La Unión; and, to a lesser extent, the Pico de Oro–Reforma and La Victoria–Nuevo Chihuahua groups.

TABLE 4.2. Priorities in the Perception of the Greatest Danger in the World Today by Community

Community	Perception			
PR (1)	Deforestation	Poverty	Attitudes	Other
VC (2)	Divine Punishment	Poverty	Deforestation	Illness
LU (3)	Poverty	Illness	Attitudes	Pollution
PL (4)	Poverty	Illness	Attitudes	Deforestation
PH (5)	Pollution	Attitudes	Environmental Degradation	Overpopulation
CR (6)	Pollution	Poverty	Attitudes	Overpopulation

Key: (1) Pico de Oro and Reforma Agraria, (2) La Victoria and Nuevo Chihuahua, (3) Lacandón and La Unión, (4) Palenque low income, (5) Palenque high income, (6) Palenque cattle ranchers.

Although they share a common concern about poverty, those in rain forest communities emphasize deforestation and divine punishment, whereas in Palenque and the surrounding areas there is a greater emphasis on pollution and illness. Most of these differences, however, are gender based, since it is the men in the forest who are most concerned about deforestation and the women in Palenque who refer to pollution, as shown in table 4.3. There is a simple explanation for this. It is the women who look after sick family members, particularly children, and the living conditions of low-income groups in the deforested area of Palenque tend to foster illness due to unsanitary conditions and a lack of hygiene, which are also results of pollution.

In comparison, there is little difference in the answers given by Indians and non-Indians. The survey showed that Indians are not worried about overpopulation and are less concerned than non-Indians about poverty, pollution, illness, and environmental degradation. Conversely, they are more concerned about attitudes.

TABLE 4.3.   Percentage Distribution of Perceptions of the Greatest Danger in the World Today by Gender

Answers	Gender		
	Total	Men	Women
Poverty	13.7	14.8	12.5
Deforestation	6.7	8.3	5.1
War[a]	23.8	25.9	21.8
Pollution	8.3	7.9	8.8
Illness	7.9	4.6	11.1
Environmental degradation	4.4	5.1	3.7
Overpopulation	3.2	3.7	2.8
Attitudes	8.3	6.9	9.7
Divine punishment	4.4	3.7	5.1
Other[b]	3.5	3.7	3.2
Don't know	15.7	15.3	16.2
Total	100.0	100.0	100.0

[a]The high number of responses regarding war as the greatest danger in the world was influenced by the fact that the survey was taken during December 1990 and January 1991, when the war in the Middle East was imminent. This shows the impact of the mass media on remote communities.

[b]Includes the lack of technology, natural disasters, the Mexican government's ban on tree felling, and the forest's natural disasters.

In Their Own Words

War

Interestingly, at the time the survey was conducted we interpreted the response "War" as a reaction to the Gulf War, but the Zapatista uprising of 1994 would now indicate that perhaps some respondents, especially women, who had knowledge of the training camps of the Zapatistas, may have been expressing their misgivings at the coming events.

One answer that exemplifies the new perception of globalization was that given by a young mother in La Unión who, having seen the signals of war on the other side of the planet, went on to say, "Now diseases come through the air, like it says in the leaflets; they come and get our children, and now we're going to get all that gunpowder."

Poverty

Most answers, in both the survey and the interviews, talked of the fear of poverty in the following terms: "The crisis is really bad. You can't support a family anymore, as far as I can tell"; "Now there's no harvest and nothing to eat." A Chol peasant farmer stated, "Life gets more difficult every day because there are a lot of peasants out of work." The words of Raúl Márquez, from the San Manuel ejido near Palenque, summarize the pessimistic view of present conditions and help explain how the Zapatistas gathered grassroots support:

> Things are going from bad to worse. Everything's going to go up, and what we earn isn't enough. And the land's stopped producing. It has to be fertilized. We don't get any help, from the government or anyone. We're all in a bad way, and the president doesn't give us any help. He helps the cattle ranchers, though. It's worse for people with children.

Poverty was emphasized even more in those communities that depend on wage labor: Lacandón, La Unión, and the Palenque low-income group. It is second in importance in all the communities in Marqués de Comillas, which are at risk from fluctuations in harvests and agricultural prices.

Pollution

When we asked further questions from those who had given the answer "pollution" in the survey in Palenque, it became clear that most of them meant air pollution, which, in fact, is not an environmental problem in Palenque. Further inquiries cleared up the mystery: they were simply repeating what they had seen on Mexican television, in which constant mention is made of air pollution in Mexico City. This corroborates the idea that "people do not live where they live" but, rather, where they are made to live by the mass media.

Other answers referred to pollution from trash and open sewers, which, together with water scarcity, is one of Palenque's visible problems. Trash is thrown into riverbeds and vacant lots, newly arrived families settle in these lots, and the pollution translates into illness, particularly among children.

Deforestation

Deforestation was a priority only in Pico de Oro and Reforma Agraria, both old settlements in the rain forest. This was probably due to the intense development projects which the PASECOP, a government program carried out in these communities at the end of the 1980s. Farmers' meetings and workshops had been held to discuss sustainable agriculture and rain forest conservation programs.

In both Pico de Oro and Reforma Agraria people generally were environmentally aware: "The greatest danger is the ending of the forest, because it acts as a lung for the world, and the oxygen would all get used up"; "We have to stop destroying the forest, it was a good thing the government put an end to the destruction." To halt deforestation, one farmer suggested; "First of all, you have to make people aware. For me the greatest danger is, if we farmers keep cutting down the forest, we'll use it all up."

Interestingly, some of them have environmental knowledge from previous situations. In Reforma Agraria respondents pointed out the danger of "destroying the forest, because it won't rain anymore. In Oaxaca there was just forest. They destroyed it, the river dried up, there were no more fish, and the land was no good anymore." They fear a similar future in Marqués de Comillas. Many have already made the link between deforestation in the Lacandona forest and global change:

"Whenever there are talks, everyone's got their eye on us. Other countries are interested in us not destroying the forest. The danger is that it will become a desert and there won't be any rain or water or anyone"; and "As for the biosphere, as human beings, we are aware we shouldn't cut down the trees in the hills, because other countries need them."

Attitudes, Illness, and Overpopulation

Attitudes were considered one of the greatest dangers in all communities. For example, some respondents stated: "Humanity itself is wicked and perverse. People invent weapons, instead of helping the poor"; "Not obeying the government about not cutting down the trees. The government doesn't allow us to cut them down"; and "Drug addiction—that's something that terrifies me."[6] A mother in Lacandón said, without the slightest hesitation: "I know what worries me—it's the bars, where our children get drunk. That's what really gets me down."

When the answers are analyzed by gender, we find that it is predominantly women who refer to attitudes as one of the main dangers. This is understandable, as gender roles have placed women in these communities at the center of moral issues and emotional relationships.

Illness was mentioned as a priority predominantly in communities with the worst pollution and the least access to medical services, including La Victoria and Nuevo Chihuahua in Marqués, Lacandón, La Unión, and among low-income groups in Palenque. Predictably, the high-income groups in Palenque did not refer to this factor.

Overpopulation was the fourth highest item of concern among high-income groups in Palenque and cattle ranchers. It was hardly recorded among the Marqués de Comillas communities and did not appear at all among the low-income groups in Palenque, El Lacandón, and La Unión. The expressions used in relation to this response and in many interviews were, for example, that "more and more people are always being born. There isn't enough room for them all"; and "We cause the danger ourselves; the more of us there are, the less work there is for everyone." In interviews this was reflected in statements such as the following by Genaro Zapata, a ranch owner:

> Here, nature is exuberant; it devours us. You can practically hear the grass growing. But we aren't able to exploit the forest; the

problem is overpopulation. We need family planning programs, because there's no more forest left.

An ejido owner from Benemérito pointed out that

> the inhabitants of Mexico have already increased so much that what you produce isn't enough yourself or for the national markets. We've been having shortages in the region, because of the number of inhabitants. When there were only sixty of us ejido owners, we didn't have this sort of crisis.

Fieldwork showed, however, a generally passive attitude toward overpopulation, as exemplified by Santos Hernández, a milkman from the San Mateo ejido, who commented: "God said that the Earth was going to get overpopulated and mankind was going to go hungry, and that's exactly what's happening." This contrasts with less frequent statements reflecting concern about taking responsibility for such processes as overpopulation. For example, in the words of a civil servant from Palenque:

> The forest has a serious problem of overpopulation, and we have to do something about it fast. Families are very big, with eight, nine, or ten children, and it's the church's fault, with all that stuff about having all the children God wants you to have.

Interestingly, while women in the region do attend mass and consider themselves religious, their views on family planning seem to be diverging more and more. María González López, the wife of an ejido farmer from Nuevo Chihuahua, simply rejects this doctrine:

> [You have to have] the children God wants you to have? That's not true. We can't bring God into this because it's a decision for each couple to make. My mother had eighteen children, but I had my tubes tied after the third. Times have changed; you can't support them [children] the same way anymore. Men want to have children because they're macho and to prove they're men, but being a man means being a responsible father, although some of them don't understand that, and then people say that a man who lets his wife have her tubes tied is an idiot because then his wife's going to be

unfaithful, and that's not true either. Couples didn't used to be like they are today; now they're a real mess. Women don't put up with things like they used to; the children are a disaster; and, when there's a problem, it's always the woman who has to sort things out.

It is striking that more men than women cite overpopulation as a current threat—but with a difference. For women the issue is personal and domestic. For men it means that more migrants will be coming into the rain forest, as is continuing to happen, and pressure on the land, both in the settlements and for agriculture, will increase.

Answers related to divine punishment implied that God would soon be punishing humans because they are sinning so badly. The high percentage of this response in Nuevo Chihuahua came from a group of Seventh Day Adventists, who typically gave such answers as: "We always study the Holy Scriptures, and they tell us of the dangers there are, but the time of anguish and much enmity will come." The response of a woman in La Unión was equally surprising: "If God decides that it's all over, that would be even more dangerous than war."

Other Answers

A few answers were given citing the main danger facing the world as the lack of technology, the possibility of natural disasters, the ban on tree felling, and the forest's natural dangers. An older woman from La Victoria provides an interesting reflection:

> In my mind I can't see anything [threatening]. In my time, when I was a little girl, there was a great shortage of corn, and now you don't see that. Before people were hungry, now they are being productive. What I have noticed, though, is that there are more plagues—that's what happening nowadays.

This reply is worth noting, since it was one of the few to assess current dangers in relation to a period in Mexico when there used to be famines. It would certainly be a pity, now that famine has been eradicated and people have land and seed, if the new threat today were extensive plagues, partly as a result of monoculture as a form of agricultural high-tech production, or the fall in food production because of soil depletion. Several people touched on these points, such as one who pointed out that "the land is no good for agriculture anymore. It's all burned up."

Only one person surveyed referred to the threat from other, more powerful nations: "As far as I can see, the greatest danger comes from the powerful countries, I mean the imperialists, as we poor go on being humiliated and exploited." The most fatalistic reply was: "Whatever happens, we can't do anything, so everything's dangerous."

Perceptions of Changes in the Natural Environment

The first difference we found is that, although only 6.7 percent of those surveyed cited deforestation as the main danger today, most of them perceived significant changes in the environment related to rain, heat, winds, floods, and the disappearance of animals, all of which are related to deforestation (see table 4.4).

Answers to questions about environmental change, however, must be dealt with skepticism, since research has found that people often say that changes have occurred when this is not borne out by empirical data (Whyte 1985:403). We have no reliable long-term data on rainfall, or other climatic phenomena, in the Palenque area and only personal or ethnographic statements that there are more droughts than before, more torrential rains, stronger winds, and more serious and frequent flooding in the forest. Therefore, we must treat these perceptions with reservation.

TABLE 4.4.   Percentage Distribution of Those Who Believe That Changes Have Occurred

Community	Rainfall	Heat	Winds	Floods	Animals
PR (1)	90	57	30	73	73
VC (2)	49	32	14	21	61
LU (3)	84	74	55	22	96
PL (4)	73	73	50	42	85
PH (5)	96	63	21	13	96
CR (6)	90	69	25	44	98
Total	78	59	33	37	82

Key: (1) Pico de Oro and Reforma Agraria, (2) La Victoria and Nuevo Chihuahua, (3) Lacandón and La Unión, (4) Palenque low income, (5) Palenque high income, (6) Palenque cattle raisers.

Which Environmental Changes Are Most Often Perceived?

Of all questions asked in the survey, the one concerning changes in the natural environment had the greatest consensus among responses. Of those surveyed 82.2 percent noted that there were fewer wild animals, 78.5 percent cited changes in the level of rainfall, 59 percent mentioned higher temperatures, 32.9 percent noted changes in wind patterns, and 36.8 percent named floods as a noticeable environmental change. A few interviewees referred to the entire range of natural changes in deistic terms such as one evangelist, who explained, "Throughout the centuries it was written that there would be changes wrought by God in the twentieth century."

Perceptions of the loss of biodiversity (including many kinds of birds, jaguars, and other wildlife) were predictably high in the already deforested areas around Palenque: more than 96 percent among high-income groups from Palenque as well as the cattle ranchers and people in Lacandón. A cattle rancher from Palenque expressed this view eloquently:

> Now they've really destroyed everything! There's nothing left. I wish I'd had a camera to record it all. A man would ride along with his shotgun, and in a single day he would come back with a deer on either side [of the horse] and a boar on top of them. Another horse would be carrying a collared peccary and a spotted cavy. Many people made their living out of that. . . . But that's all gone now.

When asked how one could put a stop to this destruction, a blank look came over the man's face, and answered wearily, "Well, that's going to be very difficult."

In Palenque illegal animal trafficking was also mentioned: "I have a piece of the reserve in my ranch, but people come in to hunt. In the beginning I had alligators, but they killed them and sold the skin, just like they did with the deer." Another explained: "People killed the animals because they [the government] didn't provide surveillance in time." Some regret this loss for economic reasons. In Palenque a waiter's wife said, "There aren't any more animals. There used to be plenty. We were really well-off because it was good, tasty meat. Now it's really expensive." A different, but interesting, remark was made by a woman from La Unión who said, "My hens and turkeys died from the heat."

In the rain forest communities, where wild animals are still to be seen, there is a widespread perception that the animals are disappearing, as shown in table 4.4. Farmers in Pico de Oro commented somewhat reticently, since they know they have been branded as the main predators. They said that the animals "had moved into the middle of the forest." Others, like the ejido farmer in Nuevo Chihuahua, were not so reserved: "The animals have disappeared. We know there's a ban, but, if we see one, bang!"—he gestures, making his hand into a pistol. Farmers in Reforma Agraria, who have undertaken measures for the animals' protection and are even experimenting with a program to protect the macaws, were cautiously optimistic: "The animals had begun to disappear, but maybe with the government ban there are going to be more."

Questioned about the rainfall, a high percentage—over 80 percent in four of the six communities—had noticed changes. Unlike the question on animals, in which there was a clear difference between those in the forest communities and those living in Palenque, there was no distinction between the two in the question on rain. For example, inhabitants of the forest communities remarked: "Now the rain is unusual. It's badly distributed"; "Before it used to rain a lot, but the rivers didn't swell"; "When the mango trees blossom, the rains come and wash them away." In Palenque people commented that "before the bad weather used to last two weeks; now it don't last very long at all."

Some of the interviewees replied that it rained more now, while many others said that it rained less now. In Nuevo Chihuahua, for example, residents replied that "it used to rain a lot, but not anymore"; "It rains less, with eight months of drought"; "It's drier. In 1986 they burned down the forest, and in 1990 there was more rain." The same contrast was recorded in other communities. The explanation for this apparent contradiction was provided by a few interviewees in Reforma Agraria: "The seasons have changed. You never know when they're going to be anymore." In La Victoria some said: "The dates of the rains have changed. This time we had bad weather in December." In La Unión the response was similar: "Before we used to have more rain, but they've spread out a bit now. Before it used to rain in June, and now it doesn't rain until August, so we sow the seeds any time during the year." And, again, in the Lacandona forest people stated, "In June we used to get sudden storms, but not anymore."

These comments clarify that what has changed is the distribution of rainfall throughout the year. It rains less heavily than usual for several months but then rains torrentially during one or two months. Thus, there is a longer period of drought and a shorter period of intense rainfall, of the kind described in García Márquez's novel *One Hundred Years of Solitude*. This double phenomenon is accurately reflected in the comment: "Either we run out of water, or it pours with rain and everything gets swept away."

As a result of these changes in rain distribution, the number of floods, especially those involving the forest's rivers, has risen. In December 1990, when we carried out our fieldwork, the Lacantún and Usumacinta Rivers had overflowed due to heavy rainfall in the Guatemalan highlands. Crops on the fertile lowlands of Pico de Oro and Reforma Agraria were swept away by the Lacantún River, which explains the high percentage of the following reply among members of these communities: "Before it used to rain a lot, but the rivers didn't rise." In the Palenque area inhabitants noted: "Now it rains an awful lot. The Chancalá River floods the motorway." Yet some of the interviewees pointed out that water storage has fallen to such an extent that, in the city of Palenque and the outlying areas, there are constant water shortages.

The response concerning higher temperatures was, not surprisingly, given most by those from the deforested area surrounding the city of Palenque (see table 4.5). Over 72 percent of respondents in the five communities in this area reported a hotter climate, with comments such as, "Before the climate was balanced, but now it's much hotter" and "The sun's rays burn more, and eight or ten years ago they didn't."

Finally, no clear pattern emerges among responses regarding changes in wind patterns. This could be interpreted to suggest that the

TABLE 4.5.   Heat Changes by Region (Percentage Distribution)

Region	Yes	No	No Response
Forest	30.7	65.6	63.6
Palenque	61.3	34.5	35.4
Total	100.0	100.0	100.0

two rural communities in Palenque (i.e., the two in which deforestation has been greatest) are the most severely affected by the winds, while changes in the winds are only dimly perceived by the other groups. What people mention most frequently is that the mango crop is not as abundant as it used to be because today's rough winds prevent trees from being pollinated. In these responses there are practically no differences in perception between men and women, mestizos and Indians, or among religious groups. In the face of clear, empirical phenomena, generalized perceptions are beginning to emerge, in which cognitive or cultural differences regarding the environment are no longer relevant. One could say that, as a consequence of ecological phenomena becoming increasingly visible and more generalized in scope, the semantic field of perception of these phenomena, which is highly divided at present, will become unified. Those in rural communities perceive these phenomena most directly, whereas in the urban community the perception is modified by television and radio.

As a last point, it should be pointed out that very few of the interviewees established a link between these changes and deforestation. Only one cattle rancher expressed the connection explicitly: "Deforestation has changed everything." As mentioned Protestants, for their part, tend to associate these changes with God's will. The majority of the interviewees, and the area's population in general, did not associate the disappearance of the forest with climatic changes or the loss of biodiversity. They associate the loss of animals with hunting for food by the peasants and with smuggling, but not with the disappearance of the forest. Conversely, the destruction of the forest is associated with the loss of firewood, timber for building, shade, and, finally, the countryside.

Chapter 5

Perceiving Others as Guilty, Vulnerable, or Responsible

Individual perceptions begin to be enunciated based either on direct observation of a phenomenon or on information transmitted by others. These perceptions are contextualized in a "subjective framework," as Anne Whyte puts it—that is, transmitted through oral traditions, schooling, or the mass media. Having identified, in chapter 4, the topics of major concerns to those living in and around the Lacandona rain forest, here and in the following chapter we focus our analysis on the contextualization of concerns in the conceptual framework of social relationships.

As explained earlier, we begin our analysis with the premise that individuals begin to create new perceptions, such as those about environmental change, by trying to fit them into previous frameworks that codify their relationships to their own and other social groups. This premise is particularly well adapted for the predominantly corporate communities in rural Mexico, as shown in previous studies carried out by the authors. We are not reifying these particular social groups; instead, we find that they are continuously being formed and reshaped while at the same time leaving a lasting impression in time. In other words, we believe that corporate affiliations will have a decisive influence on public policies and private strategies for sustainable development.

One of the most fascinating aspects of our fieldwork in the Lacandona forest was being there at the very moment at which individuals' perceptions and assessments on deforestation were beginning to crystallize into distinct positions relative to what "others" were saying or were thought to be saying—among them, the Mexican government, the international community, inhabitants of cities, and the very scientists who were conducting the research. Those of us involved in this research became keenly aware of our positions, since the people in the forest made us aware of them. These facts, in themselves, are field data. It is thus worth noting that, among the researchers, our initial positions were varied: four were strongly concerned with saving the forest, three gave priority to helping the farmers, and three had no clearly defined position.

As soon as the farmers, especially those in Marqués de Comillas, realized we had come from the national university to study "what was happening in the forest"—an expression we had agreed to use from the outset—there were various groups that immediately assumed we were *"ecologistas"* (in their definition, militants in favor of the rain forest and against the farmers). When it became clear that we were genuinely interested in finding out their points of view, they not only became extremely friendly, for which we are grateful; they also went out of their way to explain their views so that we could publish them. In fact, those who were already seeking genuine alternatives for sustainable development in the area asked for our help, so much so that our ecologist stayed on to work with them, and our agronomist and veterinarian were extended offers to work for them.

More than many other research studies, for this one anthropologists cannot remain aloof because the issue of the management of forest resources is presently an urgent one. We therefore plunged into our discussions, and, in the course of them, we could see that the distinctions between pro-forest and pro-peasant positions became blurred until it sounded like a false, purely semantic, debate; you cannot save the rain forest in the name of people while ignoring the people who live off it. A selective perception in environmental issues must give way to a more cogent, global assessment of strategies for sustainability.

In the course of our interviewing we were able to detect responses that were given to create a good impression or, on the contrary, to annoy us, but these were a minority. Most were entrusted to us conscientiously by the inhabitants of Palenque and Marqués de Comillas, and we return the compliment by recording them here, so that they can be made known to all those interested in finding environmentally and socially sustainable solutions to the disappearance of tropical rain forests.

The Round of Accusations: Who Is to Blame for Deforestation?

Since humans are all equally at risk from the irreversible changes altering the world's biogeochemical balance, and since reducing this risk means that every one of us will have to pay the cost, the first, very human, response is to look to others and to try to make them pay. Schematically, the North blames the South for overpopulation, and the South blames the North for overconsumption; governments blame ecologists for overreacting, and ecologists blame goverments for not doing enough; the rich blame the poor for pollution, while the poor blame the rich for waste; urbanites blame rural people for overusing resources, rural people blame urbanites for destroying the environment.

At one point it was widely believed that indigenous peoples, as well as women, were the best custodians of nature. But in the 1990s, both eco-Indianism and ecofeminism have found their skeptics.

Given the propinquity to reciprocate when it comes to assigning blame, it is interesting to know what is said by which groups in the Lacandona rain forest.

The Views of Peasant Women

*"It's Not True What They Say about Us Destroying the
Rain Forest"*
During the fieldwork women from farming families explained that farmers had to clear the forest because, "if they don't, what are we supposed to eat?" In the words of Macaria Rodríguez, from the Roberto Barrios ejido;

> In the news they say we have to conserve nature, and I think we do have to take care of it. The government's right to be worried about the forest, because, if they destroy the forest, they'll destroy the Earth. But the thing is that, if peasants don't sow their crops, they die. Anyway, it's not true what they say about us destroying the whole forest.

Most of them, faced with the problem of the cutting of trees, said something along the lines of, "If they're not going to let us cut down trees, they should give us an alternative" and "They should at least send a bit of money."

These women make no distinction between mestizo farmers and Indians where tree felling is concerned, but, interestingly, Estela Hernández, from Benemérito, pointed to differences according to the migrants' home states: "We haven't cleared very much because my husband's from Chiapas, and people from Chiapas aren't very ambitious. People from other states, though, came with the idea of clearing the forest."

At the same time, many peasant women candidly admitted, "We don't know why we're supposed to conserve the forest." In this they are in sharp contrast to urban women.

The Views of Urban Women

"Indians, Settlers, and Government Sawmills Have All Destroyed the Forest"
Urban women of high income or from cattle ranchers' families in Palenque were very aware of the problem of deforestation, explaining, for example, "It affects people's daily lives"; "It makes agricultural products more expensive in cities"; and "It's starting to get hotter."

Quite a few felt, like María Elena Salas, a shopkeeper, that "we are all responsible for the deterioration of the environment, especially the government, because it's so corrupt." The majority, however, agreed with Rosa María Zuñiga:

> Indians, settlers and government sawmills have destroyed the forest. That makes those of us in Palenque really sad. . . . The peasants have cleared vast amounts of land; they cut down as many trees as they want. No one can do anything about it. No one can stop the peasants. Even though the cattle ranchers are more aware of the problem, they clear the forest, too, for their ranches.

This view is echoed by a female restaurant owner:

> Indian peasants are the ones who have deforested the area, which is why the climate's changed. The Indians cut down trees so they can sow their cornfields and other crops. They're not worried whether there will be any more forest afterward. . . . Cattle raising's not the main cause of deforestation because the paddocks aren't in the rain forest; they're in special plains for cattle. . . . The solution would be to have more surveillance in the ejidos or to take the Indians somewhere else.

Some professional women in Palenque had a different explanation for the farmers' predicament. Maya Miguel pointed out for example:

"The system's wrong, and there's a lot of politics involved. . . . Peasants cut down trees to sow corn because they're hungry. They should be given alternatives so that the forest can be protected, but you can't just say, 'Stop eating.'"

The Views of Women from Low-Income Groups

"You Mustn't Turn Your House into a Wild Forest"
In contrast, women from low-income groups in Palenque are not aware of the problem of deforestation, perhaps because most are recent immigrants from rural areas. Others have a decidedly urban point of view, such as the woman who said she did not agree with planting trees in houses because that meant "turning your house into a wild forest." The views of this bricklayer's wife, however, herself a Chol Indian, were repeated in conversations with other women:

> The Lacandons [Indians] have destroyed the forest. They're the ones who cut down the trees, and then the trucks come along to buy them. The Forestry Commission is supposed to check that they don't cut down the trees but as they give them a bit of money.

This view has to do with the fact that the Lacandon Indians signed concessions with COFOLASA for timber; however, the most heavily deforested areas are certainly not those where the Lacandon Indians live. This woman's negative attitude about the Lancandons can thus be explained by the Chols' resentment about the government's having granted the Lacandons special rights over the rain forest.

The Views of Men from High-Income Groups

*"People Want to Have Cattle and Drugs Because You Earn
More that Way"*
For urban men in Palenque—office workers, professionals, merchants, and students—the problem of environmental degradation in the forest is a remote issue that scarcely affects them. Nonetheless, they realize that it is in fact being destroyed, and almost all of them have clear views on what they consider to be its causes. They mention disorganized settlements, cattle raising, ejidos, motorways, corruption in government, and poor soil management. The explanation given by one sociologist, for example, was:

Timber is one of the causes of the forest's destruction, but there are others, such as the increase of cattle raising, and, as I see it, the main reason is the poverty resulting from the lack of opportunities to make a living. So, people turn to cattle raising and drugs, because that gives more money. . . . There is demographic pressure on the land. . . . If they don't want to destroy the peasants, they should give them viable alternatives.

An ecological perspective was provided by a biologist working in the area. According to his calculations, unless something is done, the Lacandona rain forest will disappear in ten years. As he explained it:

It's not a question of importing models from countries with cool climates. There is very little support and money. The population explosion is also a problem. As long as the forest continues to offer a higher standard of living, immigration will continue. It is difficult to develop and preserve the forest without turning it into a magnet for immigrants. They'll have to involve the sects, who have a really strong influence in the area.

The Views of Government Officials

"The Government Is Really Doing Something"
The views that government officials express publicly tend to be homogeneous. They insist that the problem of deforestation affects everyone, that it will have global consequences, but that, at the same time, there is a need to assign responsibility. If the government was in fact responsible for the problem, they said, that was in the past; as a result of international pressure, the government has taken charge of the problem and is proposing real alternatives by means of its forestry programs. Government officials recognize, however, that they have not found a solution to the problem and that there is a need to raise awareness among the population. One official pointed out:

There are lots of interests in the forest. I think it would be best to take people out of there and relocate them somewhere else. . . . It's the settlers who have cleared the forest so they can sow crops and raise cattle. The government is really doing something; for the first time ever, there is now interinstitutional coordination. The rubber program is a genuine alternative.

The Views of Cattle Ranchers

"The Damn Corn Fields Are Spreading All over the Place"

Practically all the cattle ranchers we interviewed shared a common view about deforestation. They agreed that the main predators are the peasants, the Indians, the hunters, and the government, particularly the Ministry of Agrarian Reform. Unlike the farmers, who accept that they have deforested the area and try to justify their actions, none of the ranchers interviewed agreed that they themselves had deforested any land. One thing they constantly referred to was the fact that they are now leaving some forested areas on their ranch lands, as they must, since, according to recent legislation, they are obliged to set aside 10 percent of their land and restore its original vegetation.

The following account by Javier Arias summarizes the views of many cattle ranchers:

> I try not to cut down the trees on my ranch because that's my children's inheritance, but what good is it looking after my trees if the government sets a bad example by being the first to clear the land? The ejido farmers are another group who spend their lives cutting down trees and stealing our land. They've got the government in a bind, and, if [you tell me] that's not true, I'll go over to the opposition, and that's why they give them everything and they bother those of us who really want to work the land.

Others, like Don Ramiro Ordoñez, point out that

> the big fires destroyed everything. You could see them all day and all night. People didn't put up barriers to stop the fire, so the fire spread really quickly. . . . Since the 1980s the climate has really changed. It's hotter than it's ever been before—there's less rain.

According to some cattle ranchers, however, the ones to blame most are the Indians:

The greatest problem in Chiapas is the Indians who supposedly work the land, but they just put their damn cornfields all over the place. Those of us who are cattle ranchers know the problem is serious, so we've started reforesting the areas with fruit trees, not wood trees, because they get knocked down by the wind, and, besides, they need very special conditions to grow.

The Views of Farmers

"However Much the News, Propaganda, Dialogues, or
Awareness We Get Thrown at Us, We Have to Clear the Forest
for Food or Money"
Unlike the other groups, men who are farmers are on the defensive, since they know they have been blamed most for deforestation, and the majority admit their responsibility in clearing the forest. In spite of this, they do not consider themselves guilty, since they place the blame on need and poverty. The most eloquent explanation was given by Eligio Corona, from the Quiringüicharo ejido:

> We have to cut down trees to feed our families. However much the news, propaganda, dialogues, and awareness we get thrown at us, we have to cut down trees for food or money. . . . We've got the information; we understand the problem. . . . The only thing that's true is that we're living in the present, so that our children can have enough to eat and go to school so they can have a future and more awareness.

The tragedy of it all is that, to feed his children today, he has to destroy that which would give them sustenance tomorrow. This predicament does not escape the farmers, for, as one man said: "How are we supposed to plant crops? We have to cut down trees. To improve our lot, we have to harm others."

The following position, here articulated by Aaron Martínez, of Nuevo Chihuahua, was repeatedly stated among the farmers:

> The forest is being destroyed because peasants have nowhere else to work. The only place where there is still some forest left is where it's difficult to cut trees. . . . The need to survive is what makes

people cut down trees. People with education don't do that, but peasants have to be up from dawn till dusk.

Evaristo Martínez, from Flor de Cacao, explained that "it would be nice to conserve the forest, but we need to plant our cornfield for food, and no one's told us why we shouldn't plant crops."

Other peasants put the blame squarely on the rich. For many people from La Victoria, the "rich" are the ejido farmers in Pico de Oro, the nearest community, where most of the ejido land was cleared in the 1970s, long before the tree felling ban came into effect, and this has allowed some of them to raise cattle. An ejido farmer from La Victoria, one of the poorest communities in Marqués de Comillas, echoed our question:

Why have all the trees been destroyed? Well, rich people cut down all the trees for their paddocks—they can afford to clear the forest, anyway they want—but that's why it's all dry now. And then they sent us into the wild, into the desert, so there's poverty here now. It's not true what they say about not cutting down the trees and that, if we do, people will get poorer. It's the rich who cut down the trees.

Many others blame the government: "The forest is being destroyed, but that depends on the president and the government, not us. The government doesn't keep its promises; it doesn't send the seeds in time." Genaro Abraham, of Nuevo Veracruz, also sees the situation as unfair:

The government cut down the timber before and sold it, and now they tell peasants they can't cut down any more trees. You can look at that tree, but you can't cut it down. . . . It would be good if we could cut down the trees—but not the rich people or the capitalists from Mexico, who've got all they need to cut down the trees.

Eucario Ribera was unrepentant:

Men from Tabasco are *matamonte* [literally "forest killers"], and I'm from Tabasco. I came here, and they give me land. Nobody helps me. They don't tell me what to do, so I start cutting down the trees to make cornfields, and, as where I come from, they do cattle raising, pepper, and coconuts—that's what I want to do. The people from other parts do the same. . . . The [Ministry of] Agrarian Reform says don't work the land, and the Ministry of Urban Development and Ecology says you cut one tree down and we'll put you in jail. No one's going to help me get out of this mess. First me and my family are going to eat, and after that I'll look after the forest.

One farmer from Flor de Cacao, Evaristo Martínez, echoing the view of many others who wished to send their message to the wider world, said:

Ecologists and people from other countries don't know what a hard life we peasants live. We have to look after our families, and we're living in poverty. . . . They don't pay any notice to us, and then they threaten to put us in the jug if we cut down any trees. . . . We peasants have always respected the laws, but what are we going to do now in Marqués? What are they going to do? Put up military bases all over the place? Why? So they can preserve the forest and kill the poor peasants?

A few farmers mentioned that "we and the cattle ranchers are destroying the rain forest, but we don't know what use it has or what would happen if it were destroyed. The only thing we know is that the animals have gone away."

Despite the complexity of the issues, once the round of accusations is over, it is fairly clear how the blame has been apportioned, who assumes the blame and justifies it, who denies any responsibility while blaming others, and who is prepared to contribute to improving the situation. It is useful next to consider who is most vulnerable to the effects of the loss of the rain forest from the points of view of its residents.

Who Will Suffer Most If the Forest Is Destroyed?

The question of who is most vulnerable to the effects of deforestation was asked, and, significantly, there was no majority answer, as can be

TABLE 5.1.   Who Will Suffer Most if the Forest Is Destroyed? (Percentage Distribution of Replies)

Group	Number	Percentage
Peasants	171	39.6
Cattle ranchers	12	2.8
Indians	24	5.6
Government	10	2.3
City dwellers	52	12.0
International community	58	13.4
All of us	76	17.6
Don't know	22	5.1
No response	7	1.6
Total	432	100.0

seen in table 5.1. The most interesting correlations emerged, however, from a comparison of the communities. Between a third and a half of all those surveyed regarded farmers as the most vulnerable group. The one exception is the high-income group from Palenque, which said that the international community would be the most at risk, although for a number of reasons this response may be entirely spurious. Indians were also considered vulnerable, with a sharp difference between the groups: 20.8 percent of cattle ranchers and 16.7 percent of the high-income group from Palenque thought that Indians are most threatened by deforestation. This view is most prevalent especially among the cattle ranchers, professionals, government officials, and office workers: "Since they're the ones who live there, it's their environment. It's as though someone had knocked their house down."

Equivalent figures were practically insignificant for low-income groups in Palenque (8.3 percent) and for Lacandón–La Unión (2.1 percent). "The Indians live there. They don't like to cut down trees because that's what keeps it cool."

Significantly, the question of the vulnerability of Indians elicited a zero response among all the groups in Marqués de Comillas. This reflects the non-Indian settlers' resentment toward Indians because of the latters' perceived preferential treatment in the colonization of the forest.

Strangely enough, the Indian settlers—Chol, Tzeltal, Tojolabal, Tzotzil, Chinantec, and others in Marqués de Comillas—did not regard themselves as the ones who would be most affected by the destruction of the forest.[1] This fact can be interpreted in two ways. Either Indians are aware that, on the whole, they take better care of the forest and therefore do not feel as threatened by its loss or else they are not fully aware of the problem of deforestation and do not know the extent to which it may affect them. From our observations of these communities, we think that both explanations apply to different ethnic groups, but *the differences may not depend as much on the ethnic group involved as on the group's internal organization and the success of the ejidos in which they live.*

Although it was thought that farmers would be the most severely affected, most people were unable to explain *why* this was so. They gave vague replies such as, "Peasants, because they're the one's who're fighting" (a resident of Pico de Oro); "[Us,] because the forest is the only thing we've got" (a resident of Nuevo Chihuahua); "Peasants, because they're the ones who've got less money" (a cattle rancher). Very few gave more specific responses, such as: "[Us,] peasants. What will we use to cook our food?" (a resident of Nuevo Chihuahua); "The peasants, because they won't be able to find any more wood to make their furniture" (a low-income resident of Palenque); and "The peasants, because nothing gives us more than the countryside (a low-income resident of Palenque). In the survey only a handful of responses mentioned that the peasants' vulnerability was linked to deforestation, which caused more droughts and high winds.

We consider it very important that none of the responses stated what are the two greatest effects that farmers will suffer due to deforestation: *microclimatic changes* that will wreak havoc in agricultural cycles and *land erosion*. We believe this is highly significant, in that only certain kinds of information circulate about these issues and results from scientific studies take a very long time to get incorporated into the public discourse on specific environmental problems.

Conversely, many farmers feel vulnerable not because the rain forest is being destroyed, since data clearly has shown that they lack precise information on the risks that this implies for their activities, but, rather, because of the measures that are being taken to avoid its destruction. "It's not important for us farmers to preserve the forest, although it may be for the government, since the government doesn't live off the forest. If they don't want us to destroy the forest, then they should help us," explained Evaristo Hernández from the Nuevo Chihuahua ejido. It's worth mentioning that residents of this ejido, together with those of several others along the Fronteriza Road, are among the supporters of the Zapatistas.

Prophetically, another farmer added: "There's obviously going to be conflict because the government has forbidden them to go on destroying the forest. How are we supposed to sow our crops? We have to cut down trees." This remark indicates that, unless the perception that deforestation should be stopped is internalized, vulnerability is seen as the result of the immediately observable actions of other groups, government, international community, and city dwellers—in other words, as a political problem and not as a broader problem of sustainability that will directly affect them. Only when this perception is internalized through sufficient technical information and genuine debate and participation, so that it is transformed into a full understanding of its implications for individuals' lives, will the search for long-term solutions be successful.

Except for the reply citing peasants as the most vulnerable group, the remaining perceptions of who would be the most severely affected by deforestation are fairly confused. To avoid adopting a definite position in an area in which there are still no very clear positions, many people responded neutrally that it would affect "everyone," an option we were forced to add to the survey after the first day, because it was cited so often. A resident of Reforma Agraria gave this answer: "Everyone, because the oxygen would get used up, and it would be hotter." Similarly, a resident of Nuevo Chihuahua said, "we would all suffer the same because the rich live off the poor and the poor live off the rich."

Two main reasons were given for the response that city dwellers would be the most likely to be affected. The first was the threat created by "the poison from factory smoke" (a resident of La Victoria): "The forest is like a lung. When the factories come, they cut down the forests. That's what happened where I lived in Jamiltepec" (cattle ranchers from Oaxaca). The second reason was "because here we plant what is sent to the cities" (a resident from Reforma Agraria), a view that was frequently repeated, and "because the peasants are the ones who produce food" (a low-income Palenque resident).

On the subject of the vulnerability of the "international community"—the term used locally, likely introduced through workshops and brought in from the outside—it was argued, for example, that "people from other countries [are at risk] because the oxygen goes from here to there. We've got our trees, and the others can just go ahead and die" (a resident of Pico de Oro); and "The international community [is most at risk] because oxygen is given off so that people in the cities don't get so hot or tired. Mexico is the world's lung" (a resident of La Victoria). Other isolated responses were: "The cattle ranchers [would suffer most] because animals need shade to live" (a low-income resident of Palenque); "The cattle ranchers, because the grass for their animals is going to dry up" (a resident of El Lacandón); "The government, because they're the ones who buy what the peasants produce (a resident of Nuevo Chihuahua).

A low-income resident of Palenque felt that tourists would be most hurt by the loss of the forest—"because they won't have anything nice to look at anymore." The final blow was delivered by a Seventh Day Adventist from Nuevo Chihuahua, who said: "It depends on the Holy Scriptures. The disobedient shall suffer most. They shall cry out, yet there will be no mercy for them."

Who Should Be Responsible for Looking after the Forest?

We asked our respondents who should be put in charge of looking after the rain forest. Thirty-six percent of those surveyed thought that farmers should take care of the forest; 31.0 percent thought that the government should do so; a much lower percentage, 10.2 percent, thought it should be the international community, followed by 8.8 percent who thought that it should be everyone's responsibility. Practically no one, a mere 1.8 percent, thought that the cattle ranchers should take

TABLE 5.2. Vulnerability versus Responsibility (Percentage Distribution of Replies)

	Percentage	
Group	Vulnerable	Responsible
Peasants	39.6	35.6
Cattle ranchers	2.8	1.8
Indians	5.6	4.4
Government	2.3	31.0
City dwellers	12.0	1.0
International community	13.4	10.2
All of us	17.6	8.8
Don't know	5.1	6.2
No response	1.6	1.0
Total	100.0	100.0

charge, while only 1.0 percent thought it should be the city dwellers' responsibility.

A most interesting contrast can be seen between responses about vulnerability and those about responsibility, as can be seen in table 5.2. The striking feature is that, while in the case of farmers both actions are equal, for most groups responsibility for the rain forest has been shifted to the government.

An analysis of these figures shows that there are very few differences across communities regarding who should shoulder the responsibility for saving the rain forest. But there are interesting differences in terms of the reasons given. For example, a resident of Nuevo Chihuahua felt that peasants should be held responsible "because the rain forest is ours," while one of the cattle ranchers felt that "the peasants' consciousness should be raised so that they can look after the rain forest themselves."

There also were contrasts in the reasons why the government should take charge. For example, one respondent from Pico de Oro explained that "the government . . . are the only ones who make us afraid of cutting down any more trees" and a low-income Palenque resident also pointed to the government, "because of the corruption between the timber companies and the government." A Jehovah's Witness stated that it should be "the government, because they're the ones in charge."

On the basis of these perceptions people in and around the Lacandona rain forest are beginning to map out their relative positions vis-à-vis the question of deforestation.

Chapter 6

Positions and Strategies
Regarding Deforestation

Perceptions regarding the world's current dangers and the deforestation of the Lacandona forest were analyzed in previous chapters. These perceptions, set in the context of previous cultural values, and new observations are used by people in Palenque and Marqués de Comillas to create what interpretive anthropology calls new *discursive structures*. This reconstructed discourse reflects the significance that environmental transformations, and the challenges of global and local changes, hold for them.

In this chapter we chart the social map of these discourses that, through social transactions, crystallize into positions and strategies. Given the recent establishment of settlements in Marqués de Comillas, there are as yet no fixed discursive positions, handed down from generation to generation, such as those found in other traditional rural communities in Mexico. Frontier cultures evolve very rapidly.

The changing environment and social and political pressures require that positions be assumed and decisions be made in the midst of a volatile situation. Inevitably, we will be freezing in one moment in time what is otherwise an extremely fluid movement of perceptions and understandings, but this is inevitable if we want to lay bare the main trends and asymmetries that arise in discursive positions among the most important actors of the Lacandona rain forest drama.

Accordingly, for methodological purposes, the discursive positions found in fieldwork are described as though they had fixed centers and boundaries. Different positions converge around a central core, while their variations tend to merge with tangential positions, as happens spatially in any cultural universe. The axis along which positions were classified is uniquely that of attitudes about the conservation of the rain forest. This inevitably creates some overlap between positions, since, for example, those holding a dependent view may also be fatalistic in their outlook. But we chose those center points around which a significant number of interviews converged, and a significant contrast to tangential views was established. Many times the informants themselves marked those centers and those boundaries. Interestingly, some of these boundaries were enunciated as having to do with a position about the government and with a view about "development," as a theory of economic change and modernization. One view not enunciated but which nevertheless runs through different positions is the belief in the capability of the individual to change the world.

Fieldwork data consisted of over two hundred interviews as well as participant observation in collective discussions between farmers themselves and with government officials, with Indians, and with cattle ranchers. The following positions with respect to deforestation were identified: (1) dependent, (2) independent, (3) conservationist, (4) developmentalist, (5) "anti-farmer," and (6) fatalistic.

The Social Map of Discourses

The Dependent Position: "The Truth Is that the Forest Is Being Destroyed, but That's the President and the Government's Responsibility, Not Ours"

In Mexico the political alliance of peasant organizations with the state in the 1930s led to an unprecedented agricultural development in the 1940s and early 1950s, which began to decline as pro-urban policies since the end of the 1950s deteriorated the terms of exchange between the agricultural and the industrial and urban sectors.[1] As rain-fed agriculture declined severely in the 1970s and 1980s, this sector became ever more dependent on the government to eke out a living.

This helps explain why, for a majority of settlers in Marqués de Comillas, and for economic refugees from depressed agrarian regions in other parts of Mexico, everything depends on the government. Moreover, given the authoritarian tradition of the Mexican political system, everything centers on the *jefe maximo* (superior chief). A typical view expressed by those holding this position is that the president is the one who must be asked "to sort out the problems here. He is the father of us all"; and "Wherever the president goes, things improve, if only for a little while."

Even when farmers themselves are prepared to accept other options, they expect these to be provided by the government: "We don't want the government to support us, but they should at least give us some alternatives." In the words of another: "If we're just going to get all this dialogue stuff, then it'll be a waste of time. They should just give us a bit of money to keep us quiet."

Within this group quite a number, while accepting their dependence on the government, feel "deceived" by it and therefore assume an attitude of rejection and complaint: "The government hasn't done a thing. There's a lot of corruption, and they're the ones who're cutting down the forest when, supposedly, no one's allowed to fell trees."

In actual fact, there is a basis for such charges. Many informants accused the army itself and petty government officials in charge of protecting the rain forest of trafficking and of killing forest animals. One old man mentioned a case also repeated by other informants: "Two weeks ago, several *madrinas* [police vans] turned up and took away lots of animals. There's no doubt the army and police have got to give us a hand. But, if macaws and parrots are smuggled out by the army, then they should check all the Mexicans leaving." A primary schoolteacher in Palenque pointed out that "in the Usumacinta River there are turtles and the government put the army in charge of them, but they're the ones who kill and eat them. Government officials are always taking advantage." We were witness to the visit of the municipal president of Ocosingo to a remote settlement during which he demanded, in a most authoritarian manner, that they cook a meal of turtle meat and turtle eggs for him.

Blame is especially targeted to the government for the total lack of planning in the colonization of Marqués de Comillas in the early 1980s. Ernesto Mendez, from Flor de Cacao, emphatically pointed out that

in that case, if [the Ministry of Agrarian Reform] wanted people to live in the forest, they should have planned it like that. They're always changing their minds. Now that people came flooding into Marques de Comillas, they say it's a good thing that it was populated by Mexicans from all over the country.

In the ejido assembly in Nuevo Chihuahua, called especially to explain to us what their situation was, one man said that

in 1985 the government said that if we looked after the forest, they'd send us 80 tons of beans and 80 tons of oil, but it never came.

The government doesn't keep its promises. It makes demands on us but doesn't give us anything in return. If we don't cut down the forest, we won't have anything to live off. The forest is our only hope; the government is just words. If the government wants us to obey, they should keep their promises. We're right next to Guatemala. If the government doesn't pay us any notice, we'll join Fidel Castro's party.

Another farmer from Benemerito pointed out that

this government's on the way out, and the new one won't be interested in what [Governor Patrocinio González Garrido] promised, so we'll be the ones to suffer.[2] Because we're getting old. We asked for hens, pigs, sheep, and intensive cattle ranching. It's not that we don't want to work, but they gave us forty sheep per ejido, how many is that supposed to leave us each? About a quarter of a sheep. [Someone shouted, "About enough for a bit of roast mutton!"] . . . That's why we're not happy with the tiny amount they're giving us.

This position becomes more radical among those demanding that they get paid for not cutting down the trees in the forest, as summarized by the following remark: "If they don't want us to cut down the forest, they should help us." In the extreme southern corner of Marqués de Comillas, in the Boca de Chajul ejido, close to the Guatemalan border, Isidro Rodríguez echoed this position:

Now they want us to look after the forest. But who's bothered about the forest? Ranches have people to look after them but they get paid. We're not caretakers, we're *ejidatarios* [ejido farmers]. They gave us the land and said, "The land is yours now," and now they won't let us do what we want with it.

Elias Valles Orozco, from the Nuevo Chihuahua ejido, went further:

The government's worried about the forest, because it's about to be destroyed. . . . Now it's worried, because the government's connected to scientists from all over the world, with its international relations. Why did they give us the land when we asked for ejidos?

Why did they trick us? Why did they give us land that can't be worked like an ejido? The Agrarian Reform Law says that we should work the land, and now they're telling us we can't. They got us—we're nothing more than slaves.

There is widespread determination to continue clearing the forest, particularly among the communities near the border highway and the new ejidos in Marqués de Comillas. At a meeting of farmers in Flor de Cacao, Eusebio Matero was extremely forthright:

We know they've accused us of destroying the forest, but we know you have to use the forest for lots of things. They want us to plant just enough crops to survive, but we want more—not just food. I know it's not good to destroy the forest, because it gives us oxygen, and we have to take care of it. Nature's beautiful, and we need it. In the last presidential regime the army came to tell us that, if we destroyed the forest, we'd be sent to Cerro Hueco [the regional prison]. That's why we don't plant crops, but we're already dying of hunger. We need to see the government's plan for us, but before that we have to eat. That's why we're going to go on planting crops [and therefore felling trees].

An extreme version of this position demanded the use of force, as explained by Francisco, a participant at the meeting in Flor de Cacao: "We always have to find ways to pressure them [the government], because that's the only way they'll listen to us. How are we going to do that? We'll have to go up in arms. . . . The government won't open any doors for us; we'll have to force them open." This statement, obviously, can now be understood in the light of the neo-Zapatista rebellion of January 1994. Further back, in July 1991, farmers in Nuevo Chihuahua had seized several government vehicles and taken several officials hostage, demanding that the forest ban be reversed and that they be given special funding from the government.

At the same meeting, however, several others rejected this radical position, including one who answered Francisco by saying: "We have to recognize that we're illiterate and that we should cooperate. We're almost as big as a country, so we should organize things peacefully, not organize a rebellion."

Finally, a few even expressed violent, chauvinistic points of view. For example, a newspaper article by an ecologist calling for measures to stop the peasants' clearing of the Lacandona forest was met with: "That fool son-of-a-bitch, we're not the one who destroyed the Montes Azules [Biosphere Reserve] or asked for settlements. Why doesn't he come here, then? It's one thing to write and another to show your face. Ecologist wimps."

Farmers in that area did in fact invade the ecological buffer zone that had been mapped out in the center of Marqués de Comillas. When an official from the Ministry of Agrarian Reform came to ask that the land be vacated, he was met by the farmers who had invaded the land armed with pistols and shotguns. Since then the buffer zone has been removed from the maps of Marqués de Comillas.

As a footnote to the problem of moving into the Montes Azules Reserve, among others, one ejido owner explained that "no one from the Ministry of Agrarian Reform has done anything to establish the reserve's borders," to which another added, "They haven't had the guts to do it." This also applies to the government's own decrees, since the latter tend to overlap, creating conflicts and repeating the same old story of agricultural conflicts in Mexico: "Many of the president's decisions still haven't been carried out."

The Independent Position: "Politics Comes from Outside, but It Also Comes from Us"

A different position is also present among farmers in Marqués de Comillas that accentuates the farmers' own capacity for taking initiatives and their partnership with government to develop a sustainable agriculture in the rain forest. "We don't just want a monologue from the president; we want dialogue," said one of them at a farmers' workshop. Another added, "We think we're the ones who are going to have to start things off, instead of saying, 'Mister Governor, we'd like this and that.' "

This position was best reflected in the interviews with the leaders of the Julio Sabines Ejido Union. The incumbent president of the union, Pablo Gomez Santos, explained that

> Union members are aware of the need to preserve the forest, but people are worried that they won't have any means of subsistence in the forest. What will happen to our children in ten years' time? We need technology. The union doesn't want to see any more settlements or ejidos. We need education; there are no teachers

[teachers refuse to live in the rain forest], and, if there's no education, there won't be any conservation. Our children are going to do the same and set a bad example for future generations. The state government wants to train teachers in the area, so they won't leave.

Yet Gomez Santos also criticized government measures:

José López Portillo's and Miguel de la Madrid's presidential decrees contradict each other. That's why the Lacandons from Palestina invaded the [Montes Azules] reserve. They set a bad example for the other forest inhabitants. Each ejido signed an agreement forbidding the entry of any more people, but President de la Madrid and Absalon [the previous governor of Chiapas] gave them more land. If the state itself violates these agreements, how can they expect people to stop cutting down trees?

This awareness, directed toward sustainable development and the rain forest conservation, has been the result of many workshops and meetings sponsored by government programs, particularly with the oldest ejidos in Marqués de Comillas. In the mid-1980s the Interministerial Commission for the Lacandona Forest implemented a participatory program (later called PASECOP) that, in addition to providing input for production, also fostered a dialogue with the farmers through meetings and workshops. As a result, an independent position grew among the farmers of the area of the Lacantún River.

According to those who hold this independent position: "Those who take least care of the forest are the ones who have come in from Guerrero, Michoacan, and Guatemala. They asked for land, because they wanted to extract what was left of the forest's wealth, and they destroyed it." This is important, because data showed that perceptions and behavior varied according to each settler's initial motives for coming to Marqués de Comillas. Julian Sotero, from Pico de Oro, explained that "we came here with the idea of being farmers, sowing corn, beans, and chili, and driving a tractor. That's what I'm used to. And I didn't come here with the idea of selling wood, because I don't know anything about that." Since his aim, therefore, is to go on living there, he is interested in finding methods of sustainable farming and cattle ranching.

In the Reforma Agraria ejido, a tiny paradise on the banks of the Lacantún River, the whole community is tuned into finding solutions toward a sustainable development. Many share the views of Luis Hernández Dávila, head of one of the main families and former president of the Julio Sabines Ejido Union:

> Unfortunately, you can't live off awareness. People who don't have any [cleared land] will have to be given alternatives and 12,500 acres of *jimbales* [secondary growth after land has been deforested]. What produces more oxygen, *jimbal* bushes or a banana plantation? What's the difference? So why don't they use the *jimbales*? They could negotiate that, but it would have to be negotiated ejido by ejido.

Because it gives an excellent summary of the independent position, extracts are reproduced of a manifesto published in the national newspapers by the ARIC (Rural Development Association set up by the government) in the Canadas region of the rain forest, to the east of Marqués de Comillas, in response to the environmentalists' accusations that the farmers are destroying the rain forest:

> 1. We share your concern and think it is justified as regards the Montes Azules [Reserve]. We resent being portrayed as savages or depraved persons trying to create rights and violating decrees and laws. . . .
> 2. We do not regard ourselves or the Indian peasants as plunderers. The real plunderers are the cattle ranchers who clear the rain forest for grazing. The rain forest was given to us to protect the large landed estates and to settle agrarian conflicts by the Department of Agrarian Affairs and Settlement. The presidential decrees did nothing to protect the forest. The first decree was issued so that COFALASA [the government company] could log in an area of 1,535,000 acres; the second was to legalize fifteen Tzeltal and Chol settlements. . . . The forest has been used for political purposes. It was used to attract an inexperienced heterogeneous population of settlers to an area along the border, who did nothing but sell wood illegally, and then Pemex moved in. . . . Thus, the current destruction is not due to the farmers' or Indians' perverse plundering but to nonenvironmental policies.
> 3. The decrees did not plan complementary measures to defend the ecosystems: one decree was superimposed on another.

All the farmers having this position ask is that their efforts at developing sustainable agriculture and agroforestry be supported.

Farmers having this position also tend to take a global view of the problem of deforestation, but in their own terms. Artemio Benítez, from the Reforma Agraria ejido, put it this way:

> The changes in the rain forest are going to have world consequences. Something's going to happen all over the world—that's why everyone's worried. The gringos [meaning both Americans and Europeans] and the Japanese are like gods; they pay the Mexican government not to destroy the forest. They're climbing higher and higher; they've almost reached the gods. They've destroyed everything; they've got nothing left in their country—that's why they want to preserve this.

The Conservationist Position: "When People Clear the Forest to Sow Crops or for Cattle Raising, They Lose Far More than They Gain"

So begins an account by Juan Manuel Adame, a professional from Palenque, illustrating the ecological position in relation to the forest's destruction. Adame went on to say that "people at all levels should stop eating meat, especially in the cities. For example, 5 square yards of rain forest are destroyed for a McDonalds hamburger because that's what it takes to feed a cow. Meat is a luxury. Man is not naturally carnivorous." Adame ended by suggesting that changes in the natural environment should be prevented before they occur: "It's typical of Mexicans to say, 'Well, it's already happened; there's nothing we can do about it,'" and then said he was unaware of any preventive measures that had been taken to protect the forest.

This last point is significant because it suggests closed information circuits. In Palenque people are very aware of pollution in Mexico City, but many do not know of the tree felling ban for the forest, which is the most significant initiative to date and one with great local and regional repercussions. It also indicates a lack of systematic and comprehensive social communication by both the federal and state government about its policies and programs.

What we found is that radio and television increasingly determine what people think about environmental issues. As Guadalupe Cabrera, a stonemason's wife from Palenque, remarked, "I think I heard on the radio that the rain forest is being destroyed. If they go on cutting down trees, it's going to get hotter, and that might stop the tourists coming, as that's what they like about it here."

Farmers supporting conservation expressed similar views to those of Juana María Ramírez, the owner of a restaurant on the highway to Chancalá: "It's very important to save what's left of the rain forest because, once they destroy it, they destroy the atmosphere too, and it'll get really hot. It's cool in the forest, like it normally is in the mountains."

Young people, children of the ejido farmers, are beginning to show greater interest in conservation and great concern for the consequences of deforestation, yet they have no means of taking part in measures to prevent it, so pessimism and the desire to avoid the problem are widespread. This is clear from an interview with Pedro, the son of an ejido farmer from San Manuel:

> If the rain forest is destroyed, we won't be able to breathe; it won't rain anymore, and the corn will dry up. People will die of hunger if they don't clear the forest, so they should only work 2 or 3 acres because, if things go on like this, there will be no forest left in ten or fifteen years and nowhere to live or eat. That's why I'm going to study so I can go and work somewhere else. I'm going to be a teacher or an engineer. . . . you can tell a tiger by its skin. All the forest animals and all the plants are going to be destroyed, everything's going to end up like a desert. It's really sad.

The Developmentalist Position: "When There's Electricity and Television and Paved Streets, It's Going to Be Really Nice around Here"

The developmentalist position is held by those who want to have highways and urbanization in the rain forest settlements. "Five acres is enough to live off, but what about people's progress and development? We need roads to get things out of here," said an ejido owner from Boca de Chajul.

In an ejido on the border with Guatemala, Don Aristeo Menendez, the ejido commissioner, whose interview was typical of those we heard, suggested that building a paved highway through the forest to Guatemala to turn the ejido into a commercial center should be of the highest priority:

The ejidos on the border are going to be very good for business. . . .
People from Guatemala, Central America, and Costa Rica are going
to buy dairy products and toothpaste from us, because everything's cheaper here—maize flour, gasoline. But they need someone with money to set up an exchange house; that's what we need
to get trade started.

Menendez insisted on the need to produce wealth by cultivating
extensively: "[To do this] we need machines, tractors, to cut down trees,
more technology, and fertilizers for large-scale agriculture, like they
have in Sonora [he himself had been a tractor driver there]." His
developmentalist vision is categorical. When Veronica, the research
team's ecologist, suggested that some of the techniques used by the
Indians living in the Lacandona rain forest might be learned and
adapted, Don Aristeo answered haughtily: "Unfortunately, they're very
primitive people, and we're not prepared to live like that." On another
occasion he went on to say:

Most people in the ejidos are uncivilized; they can't even see beyond a year. And you can take my word for it, the man or leader
who educates his people and sees into the future, the men who see
beyond their own noses, they're the ones who prosper; if not, they
get stuck and never grow.

The interesting point is that Don Aristeo is the prototype of a rural
leader—part chieftain, part father—who has a vision of the future and
with whom it is necessary to work because he is capable of mobilizing
the community for development. Yet he refused to budge an inch in his
commercial and urbanized vision of the future for the rain forest. Later
on he was asked by Licha, the research team's agronomist: "But don't
you think, Don Aristeo, that tractors are going to make the soil compact
so the water can no longer flow and plants will not flourish?" which is,
in fact, a technically correct observation for the rain forest soil. He
answered: "We have to cultivate the land here. We have to produce
more corn for everyone and to send to other cities. We have to cultivate
the land to create wealth." "And what about after you've ruined it all
with such cultivating?" Licha insisted. "Afterward, we'll just sit down
and cry," answered Don Aristeo.

*The Anti-Farmer Position: "No One Can Do Anything about
the Farmers; They Just Can't Stop Them"*

The owner of a handicraft store in Palenque summarized this position:

> Cutting down the trees won't do the farmers any good in the long
> run. In the short run, they get land, they cut down the forest, they
> get credits, and then the forest burns down and it's all over, and the
> taxes people pay go down the drain, and the country goes to the
> dogs. It's not worth cutting down the forest to make ejidos; they
> should've given them something else, not the forest. They need
> technology and training. They should send people to Japan to learn
> how to be productive with just a tiny bit of land and make sure that
> farmers stick to their plot of land, but, if no one goes there, no one's
> going to supervise them. For example, a rich man from Chancala
> had 125 acres of virgin forest, but the ejido farmers moved in and
> burned down all the trees. They've got no initiative. They don't
> produce anything, so they have to buy everything in Palenque.
> Those trees at the entrance to the ruins are just a facade; half a mile
> ahead it's all paddocks.

Among the farmers there are some who criticize other farmers'
actions, such as the woman from Benemerito who said, "Sometimes, the
farmers have their own land, and they just burn down trees for the heck
of it."

In a more extreme form a housewife said:

> Palenque isn't like it used to be. There's an awful lot of tree felling
> going on. The farmers have cleared huge amounts of land. They
> ask permission to clear 25 acres, and, since they set fire to the trees,
> they burn a lot more, so they cut down as many trees as they like.
> No one can do anything about the peasants. They just can't stop
> them."

A cattle rancher also remarked: "You can imagine how much money
those Lacandón bastards and ejido farmers who got the rain forest
made. . . . I'd like to be a rain forest ejido farmer so I could sell my
wood and earn money doing no work."

This view is generally accompanied by despair at the inability of
society and those involved to solve the problem of deforestation.

Rodrigo Mateo, a primary schoolteacher in Palenque, is one of many who think that "the government should set up programs, because, if they don't, nobody will. You have to raise people's consciousness, but so far I haven't seen a single program to do this."

There is a link between this position and the dependent one. Both appeal to the government to resolve the local situation, the former to stop the plunderers and the latter to ask the government to give them everything.

The Fatalistic Position: "I Think That the Forest's Destruction Is the Fulfillment of a Prophecy . . ."

Some people in the Lacandona region believe that saving the rain forest does not depend on human will but, instead, on supernatural powers. This is the fatalistic view, whose roots, according to traditional literature on Mexican fatalism, are linked both to Mesoamerican culture and to Catholic doctrine. Yet, looking at more contemporary events, passivity is also the result of a long-standing political regime of paternalism and authoritarianism in rural Mexico.

The first observation to be made is that all those holding a fatalistic view are not necessarily indifferent to the fate of the rain forest. For example, a Tzeltal farmer from the San Miguel ejido, David López, said: "I don't know what we're going to do about the forest. It'll last until it's destroyed, maybe another fifty years. They're destroying the forest and its beauty, too." The salient point is that, although López appreciates the forest's beauty, he feels totally powerless to prevent its destruction.

It could easily be said that this is a common position among Catholics, yet it is important to note that Chiapas has had a strong Theology of Liberation movement, led by Bishop Samuel Ruiz of San Cristóbal de las Casas, and priests have been especially active in some regions of the rain forest, preaching a new doctrine of activism. This activism, however, has centered on the struggle against poverty and injustice and has totally ignored enviromental issues. The activities of these priests were a major factor in the neo-Zapatista uprising of 1994.

Opinions differ, however, about their activities, as Emiliano Martínez, an ejido farmer from Nuevo Guerrero remarks:

I'm not Protestant or Catholic either. The priests are real trouble-makers. They defend the lazy ones who want to scrounge off the forest. Let's all feel sorry for the poor. Those priests should go to the whorehouses to watch the poor gambling away their money and screwing any woman they can get their hands on. If they call themselves priests, they should make the poor stop drinking—that's why they're so poor in the first place. I've yet to see Bishop [name omitted] tell them to stop drinking and set to work and stop beating their wives. Otherwise, what good is the church, if it just encourages their vices, don't you think?

It must be explained that the Bible itself is cited both for environmental concerns and for giving priority to human concerns over environmental ones. Like all holy scriptures, it is a source of infinite exegeses that can be interpreted in a variety of ways.

A further complication is the rapid expansion of Protestant and other sects in the region. They are the ones who most profusely quote the Bible. One woman interviewed, the wife of a government employee, had this to say about them:

> [the sects are] all mouth and no ears because they think they know everything from the Bible, and they don't listen to anything. They're fundamentalist sects financed by the United States who started coming here in 1980 when the conflict in Central America was at its height. They manipulate people and spend their time praying, and then we get lots of North Americans coming in with their well-equipped trucks.

Although they have recourse to exegeses from the same Bible, Catholics tend to show passive attitudes toward environmental phenomena, while the sects express more active points of view, although they still leave the final decisions to God.

This was borne out by an experience during our fieldwork in an interview with a group of Seventh Day Adventists. As they continued to work on the palm roof of their new temple, we asked them what would become of the forest in twenty years' time.

> "In twenty years' time?" mused one of them, "That's the year 2011. Well, we won't be around anymore."
>   "Why not?"
>   "Because the world is going to end in the year 2000."

"How do you know?"

"Because that's what it says in the Bible."

"In that case, why are you building a new church?"

"Well, because we've still got nine years left."

The Jehovah's Witnesses' attitude toward deforestation, which is similar to that of other sects, can be summarized in the words of Eze-quiel Figueroa, a preacher from this sect, who declared, "Man is destroying the forest for his own benefit, that's not right, even though he's also doing it out of necessity, because Mexico is a Third World country, where there's a lot of poverty."

Fatalism also blends with apocalypse: "I think that those places in the forest must have been for pleasure, because God wants evil to be destroyed, and may His will be done."

An Extreme Position: "You Need to Be a Dictator to Save the Rain Forest"

A number of people fear that current measures will not suffice to pre-serve the rain forest and that a series of confrontations will occur, as the neo-Zapatista uprising gave witness to. This fear and the skepticism behind it have produced categorical positions on the subject that are idiosyncratic to some, realistic to others. For example, according to Moisés Montero:

> You need to be a dictator to save the rain forest and kill the bastards who are destroying it. The agronomists are all corrupt. You either need the international community to be in charge of saving it, because the government doesn't work, or a dictatorship. Experi-ence has turned me cynical and made me talk like this.

Survey Data on These Positions

How many hold these various positions, and who are they? The survey data provides an approximation. First, it reports more pessimism than optimism about the fate of the rain forest, as can be seen in table 6.1. Almost half of those surveyed think that the forest will disappear, end-ing up "like the Sahara," according to one. About a third think the forest will be preserved, but 22 percent think that its conservation will depend on several factors.

TABLE 6.1.   Replies to the Question "What Do You Think Will Happen to the Rain Forest in Twenty Years' Time?"

Reply	Number	Percentage
It will be preserved	39	9.0
It will be preserved if . . .	95	22.0
It won't be preserved	211	48.8
Other	57	13.2
Don't know	30	6.9
Total	432	100.0

In order of frequency the forest's conservation was thought to depend on: (1) government prevention of tree felling: "If we take care of it the way the Ministry has told us to, we'll survive; if not, we won't last long" (a resident of Pico de Oro); (2) providing alternatives for the farmers: "Many farmers will start destroying the forest because money doesn't go so far these days and that's when the trouble's going to start" (a resident of Nuevo Chihuahua); (3) raising people's consciousness and enlisting their cooperation to preserve the forest: "If we go on educating children, they'll have a better idea of how to look after the forest" (cattle ranchers); (4) preventing an increase in the forest population: "If the population goes on rising, things are going to get worse and worse. The government should start getting people out of there" (a resident of Pico de Oro) and "The forest is full of people who have lots of children and that'll be the end of them" (a resident of Palenque Altos); (5) curbing the expansion of cattle ranching: "People are going to go on destroying trees; unless the government protects them, they're all going to be turned into grazing grounds" (a resident of La Unión); (6) stopping oil company Pemex from building any more highways or oil wells: "If they don't put a stop to things, everything will be destroyed; the highways and Pemex destroy everything (a member of Palenque's high-income group).

A comparison of these responses revealed no significant differences between men and women, Indians and non-Indians, or members of different religions. On the other hand, responses did vary significantly according to the different regions and communities. In the forest region 88.2 percent of those surveyed replied that the rain forest would be conserved, as opposed to 11.8 percent from the Palenque region. Obviously, the forest has already been destroyed near the latter, so Palenque's inhabitants are not very optimistic that deforestation in the rest of the area can be curbed. Besides, most people there have never even seen the rain forests, even though it is only five hours' drive from Palenque. In the words of Mrs. Uribe from El Lacandón: "Well, we really don't know the rain forest, although we sometimes see it on television."

On the other hand, 36.5 percent of the forest inhabitants envisage a different future for the region that will have nothing to do with the forest, a view that is held by a far higher percentage in Palenque. This shows that people in this region no longer imagine a future associated with the forest, even after reforestation. Inhabitants of the two communities in already deforested areas, La Unión and El Lacandón, envisage a particularly bleak future:

> The way things are going, things can only get worse. Everything's going to be more expensive. Now, if you plant something, you have to take very good care of it. There are so many pests, and the animals get in everywhere. For example, bananas used to grow really well, and now their roots get all worm eaten. It's going to get worse in the future, with everything so expensive, and, as for those of us without any money, how are we going to manage? They give the landowners credit, and, as for the government, we're not sure if the bureaucrats keep the money they're sent. Things seem to get worse every day, because people are very divided; there's no union. There are ignorant people who want everything for themselves and nothing for others, and, in addition, they want to get rid of them. There may even be killings, because of all the parties that want to get involved.

The survey also showed a surprisingly high percentage of those who agree with government measures to preserve the forest, despite criticisms expressed in several accounts; this can be explained by analyzing the results by community. The following answers were routinely given: "As long as the government goes on being strict, [the rain forest] will continue to be the apple of the world's eye, but, if it lets up, the forest will be destroyed" (a resident of Pico de Oro–Reforma Agraria) and "If the government goes on doing what it's doing, the forest will be preserved, and that's good. You have to live in harmony with nature" (a high-income resident of Palenque).

But criticism follows: "I don't know what the government's plans are. They want the forest to be preserved, but they should solve our problem first. If they don't support us, we'll start destroying" (a resident of La Victoria) and "No one's going to survive. There won't be any mountains left if the government doesn't support us. We'll cut down trees, and the city people will die, like us, because they won't have any oxygen" (a resident of Nuevo Chihuahua).

The social map of positions and strategies shifts continuously, since it responds to constant *transactions* between the groups, as the accounts themselves show. The more extreme the perceptions between the groups become, the more difficult transactions will become. By way of example, the more that farmers perceive that deforestation is not a problem for them but one of interest only to the government or the international community, the more radical their demands will become. The more cattle ranchers perceive farmers as "scroungers," the more unilateral will be their demands to have the farmers controlled.

It is worth noting that the recent neo-Zapatista uprising, with its enormous mass media coverage, has indeed polarized positions in Chiapas as a whole and in the rain forest regions in particular. Environmental conservation, then, will necessarily have to involve political, social, and cultural settlements of conflicts.

Conclusion

The salient result of our analysis is that environmental change cannot be studied only as a direct relationship of an individual to the natural environment. Instead, individuals' choices and behavior toward nature are shaped and channeled by preexisting conceptual frameworks and by the matrix of social relationships in which each individual's group is embedded. In studying the Human Dimensions of Global Environmental Change, then, the first task is to identify the driving forces for such phenomena, but, second, the main processes channeling social behavior have to be mapped out. In our study we began the second task by looking at the culturally bound discourses that are shaping peoples' new perceptions and positions in relation to deforestation in the Lacandona rain forest, with the following results.

We found, first, that these new perceptions and positions are being defined according to *previously defined semantic boundaries* with other social groups, which are beginning to produce a new social map of these perceptions. Second, we found that in the majority of cases, given the lack of in depth and reliable information, *the perception of ecological changes has no intrinsic content* but is, instead, superimposed on positions held previously in a context of existing sociopolitical relationships. The only real innovation in conceptual terms is that of defining a new relationship between the local and the global so that people are beginning to use new concepts of "globalization" and "international community" and are situating themselves within this new context.

Perceptual Boundaries

If there is no center, what are the main boundaries of discourses on perceptions and positions found in this study of the Lacandona rain forest? One outstanding feature was the forest dwellers' perplexity that a local phenomenon, deforestation, had suddenly become a global problem; their region had moved from the "end of the world" to the "eye of the hurricane." Fieldwork showed the simultaneous coexistence of the different stages through which new awareness is evolving, accurately reflecting what is going on internationally.

An initial reaction by many has been to deny that deforestation is a priority issue. Rather, people are concerned, in order of priority, about poverty, unemployment, land scarcity, armed conflict, pollution, wrong attitudes, and health, more than about deforestation, and the survey analysis showed which communities are mainly concerned about which problems. It is interesting that, though conservation has been talked about for many decades in the region, has been practiced by the Lacandóns for centuries, and has been driven by pioneers such as Gertrude Duby and Manuel Alvarez del Toro, it has never before become a public issue as it is now.

In a second stage deforestation begins to be perceived as a loss, but measures designed to protect the rain forest may be seen as a threat to present survival or potential benefits. In families with no surplus money such a view is intrinsic to their situation. Thus, strong opposition to such measures arises. Those who make a profit out of clandestine logging will also be fostering such opposition.

The third stage occurs when more information shows that deforestation also constitutes a problem for the region, due to microclimatic variations, soil erosion, and flooding. The question is, then, Who should take on the cost of *redressing* this situation? Options are then discussed in terms of the costs and benefits to different social groups already enmeshed in unequal relationships.

Finally, the fourth stage consists of the active search for agricultural and urbanization programs compatible with the preservation of the natural environment. In the Lacandona region such groups are trying to widen their range of action with the support of environmentalist non-governmental organizations (NGOs), government programs, and international funding sources.

Deforestation Is Not Perceived as a Problem

Why is deforestation not perceived as a problem by many people in the Lacandona region? Primarily because the issue has come from the outside and because no detailed studies or reliable data exist on its negative effects on agriculture, health, and other main concerns of local inhabitants. But the most important fact is that these effects can already be seen in certain parts of the region, yet no social awareness about them has been created locally.

The clearest example of this is provided by the data on La Unión and El Lacandón, two of the communities surveyed, whose situation can be termed both a social and an environmental disaster. These settlements, established by government and Alianza para el Progreso programs in the 1960s, were alloted rain forest lands and left to their fortunes. The land, unsuitable for agriculture, was immediately cleared, and timber dealers and cattle ranchers finished off what was left of the forest. These actions, combined with population growth, have produced communities that today have severe problems of soil erosion and infertility, water and fuelwood scarcity, high unemployment, and no outlook for development because of demographic pressure on the lands and "*minifundismo.*" Although many have a clear perception of this, discussions about it usually center on who is to blame—the government, the farmers, or the cattle ranchers—and no action nor position is taken to stop it from happening again, mainly because it is known that decisions about development programs always come from the central government. Very few associate this social and environmental disaster with what might happen again in the Lacandona rain forest. Faced with no prospects in this destroyed area, the children of the families in these communities are now trying to move to the Marqués de Comillas region, where the cycle might simply be repeated. Why can't this cycle be stopped?

First, as the analysis of the basic concepts of nature in the communities has shown, in the traditional Catholic or lay culture no specific value is attached to the natural world. On the contrary, this value is present in the local Indian traditions, yet human behavior is passive in this respect. Thus, no endogenous social perceptions of encroaching environmental degradation have emerged.

Second, in the case of La Unión and El Lacandón nothing has been learned from this disastrous situation, because the same driving forces continue to operate. On the one hand, centralized planning made the communities completely dependent on government, with little margin of action to find local solutions to problems. On the other, population growth continued unchecked so that land allotments soon became insufficient to maintain the population. As a result, given the rigidity of discourses on the relationship with government, the communities reverted back to the same demands as in the 1960s, that new rain forest lands be redistributed—except that today the government is no longer going along with this solution.

The problem is that, as long as the general perception still persists that the rain forest is "unproductive," as one cattle rancher put it, it will continue to be seen as an agricultural frontier by communities with high population growth. If the government does not cave in to this demand, then an unauthorized settling and logging of the rain forest will continue—unless, as one informant put it, the government can put a soldier behind every tree.

The illusion, then, continues that there is still more land that can be used for cultivation. In our view the limits of this process have already been reached. Today there is a need to reverse this perception in order to emphasize the conservation of already cultivated land. Thus, it is essential that conservation of cultivated land be linked to concern for irrational deforestation and given the same degree of priority. This implies, among other things, stopping land erosion, curbing population growth, and preventing city expansion over the richest agricultural lands that in turn pushes landless farmers to rain forest regions.

Finally, deforestation tends not to be perceived as a problem because ecological sustainability is not socially perceived as the long-term basis for the economy of rural communities. Loss of moisture in the land; the fact that it rains fewer days but more heavily, leading to longer periods of drought and flooding; the disappearance of essential firewood for households; soil erosion; and the imbalance of biodiversity are all results of deforestation that directly affect agricultural activities and domestic economies. At best, these phenomena are perceived as isolated events, not linked to ecological and economic systems. Even when certain members of these communities see the connection, their economic situation is so precarious and the lack of viable alternatives so marked that they cannot halt this process.

General information and accurate data on these processes could be provided by the media and schools, but they do not do enough. Particularly acute is the problem of centralized information. Some people from Palenque regard air pollution rather than deforestation as a major problem, even though the former does not exist in their town; it is, rather, the problem that Mexican television talks about daily.

*Deforestation Is Seen as a Problem by People
on the Outside*

When awareness about the dangers of deforestation comes from the outside, given the selective local perception, the issue tends to be regarded as "a problem for the people from outside," as the survey shows—that is, people in Mexico City have no oxygen; the industrial countries have all done away with their forests (they still have no information on the fact that industrialized countries want tropical rain forests to continue to be sinks for their uncurbed greenhouse gas emissions). Locally, then, the problem is perceived as one of "pressure from the international community" on the federal and state governments to stop deforestation. This perception overlaps with the vested interests of various groups, which must be clearly distinguished.

One group is that of a minority of rain forest settlers whose main interest in getting 125 acres of rain forest land was to sell off the timber as soon as possible and then leave. Many of these are urban people from nearby towns in alliance with clandestine timber dealers and corrupt government officials. Of course, they are the strongest pressure group against government measures to save the rain forest, but they hide behind the legitimate fears of small landholders and foster their rebellion.

This group must not be confused with another group of settlers in the rain forest who want to plant and harvest crops but did not manage to clear their land before the ban against tree felling came into effect. They tend to take on a dependent or radical position in demanding that either the government or the international community be the "custodians" of the rain forest.

Some of these farmers form alliances with timber dealers who wish to continue an irrational exploitation without reforestation or with cattle ranchers or ejido farmers who want to go on expanding their lucrative cattle raising frontier. They easily become followers of political leaders who, because of their own interests, outline the problem in terms of an exacerbated nationalism that defends the right to exploit the country's resources, albeit destructively.

It should also be mentioned that this position uses corporatism as an alibi to deny individual responsibility, as could be seen in the rounds of accusations. It also justifies wrongdoing on the grounds that others—

in this case, the government—do not give what is needed, thereby placing the responsibility on other groups once again. Yet feelings about who should be responsible for protecting the forest were very clear. The majority thought the government should take charge, although most farmers were prepared to take on the responsibility, provided the government could give credit and technical advisory services to enable them to do so.

Finally, it must be said that some local groups, as fieldwork data showed, are increasingly aware of the risks of deforestation for local life. Leaders of farmers' ejido unions, government officials, and medium- and high-income groups in Palenque all agree about the urgency of curbing deforestation.

Patterns of Perceptions on Deforestation

Deism is prevalent among those interviewed, but on the question of environmental change we found the issue is not so much about how the world was created but, rather, about the responsibility humans have toward the world. In this respect the conceptual divide lies not between religion and laicism but between those who, regardless of whether or not they are religious, place the responsibility of looking after the natural world on humankind.

Where religiosity does have a bearing is on attitudes toward this responsibility. We found that Catholics, even those who assumed this responsibility, tend to act passively toward it. Protestants and members of other sects tend not to assume it, because of their adherence to the Scriptures, but, when they do so, they respond more actively. The nonreligious take on the responsibility on principle and are the ones who carry out and demand action most insistently.

Fieldwork revealed significant differences in the relationship between mestizos and Indians with nature. The latter, even those from areas with different types of ecosystems, tended to preserve more trees on their plots of land and family orchards and to clear away less forest in their agricultural plots. Their discourse on this, however, is exactly the same as that of other ejido farmers. Just as they experience the same risks of poverty and economic fluctuation of farming, they share the same ambivalence about deforestation. Like farmers, Indians state that their main concern is having enough to eat and avoiding poverty and unemployment, so that, although they regard deforestation as a problem, they do not see it as their main problem. Consequently, they justify clearing the forest for economic reasons and express a dependent view. At least one indigenous group, mainly Chinantecs, however, have wholeheartedly taken up the search for a diversified sustainable agriculture in the rain forest.

No significant differences in perceptions or positions related to deforestation exist between men and women, although different emphases can be detected. While men regard poverty and deforestation as the greatest imminent risks, women focus on illness, wrong attitudes, and pollution. Any program for the sustainable management of natural resources directed at either group must therefore approach the issue differently. The study showed that, of all the groups, female urban professionals, entrepreneurs, and employees are the most sensitive to the need to protect the environment, both natural and social. They often hold different views from their husbands, who are directly responsible for economic interests that might be affected by actions to protect the forest or who belong to corporate groups that urge them to adopt group positions. Women do not tend to participate actively in environmental projects, however, probably because the local political atmosphere does not encourage them to become organized and participate and because there is no tradition of forming voluntary associations for specific activities.

Sustainability and Cultural Values

We began by asking whether sustainability is an attribute of culture, and, if anything, the study showed that it cannot be contained only in values related to the natural environment. In other words, we found that values leading to sustainability are systemic, so that, to save the rain forest, you have to change values enmeshed with other sets of cultural values. One does not simply have to introduce an appreciation of the rain forest; rather, one has to make sure that the relative weight given to this value supersede others and is presented in a form compatible with other values.

Although we were only able to detect this mechanism and not follow it through, an example will show what we mean. Protecting the rain forest, even by those willing to accept its value, is overridden by the indiscriminate search for profit, which is a main source of social and political prestige. In turn, it is socially accepted that this profit be used for unproductive consumption: machismo converts the dispendious use of wealth into the means of affirming virility; corruption in government makes it necessary to pay bribes to get even the most minimal services provided; the amount of money spent on alcohol in Chiapas as well as in other rural areas of Mexico is disproportionately high related to rural incomes; and so on.

Many people in the region talked about the fortunes that some men had amassed in the 1980s from logging in the Lacandona rain forest, and they all asked where it had gone. Only a few of these loggers, as confirmed by other data, had invested their money in ranching or building a house. But the rest, according to informants, had spent it on sumptuary goods—on alcohol, gambling, and prostitution. This is not a problem of poverty, environment, or politics; it is an ethical and cultural one.

Consequently, in solving a problem such as deforestation, it is not enough to increase awareness by introducing a new value of sustainability; it is also necessary to adjust the values of profit and conspicuous consumption as standards of social prestige. One cannot expect limits on the irrational, individual exploitation of natural resources to become solely the government's responsibility, since this would lead to increasingly repressive measures. Therefore, those who wish to continue this irrational type of exploitation will continue to encourage the perception that deforestation is not a problem for local inhabitants so as to provoke a confrontation between the farmers and the government.

To avoid this conflict in the short and long term, there is a need, as several pointed out in the survey and in our fieldwork, for the whole of society to assume that it is in the common interest to preserve the rain forest, and the natural and social environment in general, in a sustainable way.

From Local to Global

We began these conclusions by pointing out the Lacandona forest inhabitants' perplexity at suddenly finding themselves at the center of a worldwide predicament. There is no doubt that global change is transforming all local inhabitants into global citizens. This is a new mentality that has only just begun to emerge and one that affects the Lacandona rain forest's inhabitants as well as those of Mexico City, New York, Tokyo, and Calcutta alike.

The forging of global forms of governance and citizenry could have two possible outcomes: it could either reconfirm inequalities both nationally and internationally so as to continue to favor the rich nations, or it could become a charter to argue that, just as in any nation, blatant inequalities in standards of life are unsustainable for all at a global level.

For this we need science. It needs to create a new research program, in the strict sense of a paradigm, on global change, understood as both a natural and a social process. On the human dimensions of global change it requires not only the definition of new topics and experimentation with new methodologies but also the formulation of new theoretical premises, which will inevitably lead to reopening an ethical and philosophical debate on the purpose of contemporary societies. Science can either act as a mere spectator to this debate, restricting itself to empirically showing how processes—apocalypse or salvation—go forward, or it can open itself up to this debate, becoming involved in it and venturing to be far more intellectually creative than it has been in the past few decades. In our view the young, to whom this book is dedicated, demand and are prepared to participate in this debate to construct a lasting future.

The book has shown that there are rain forest farmers, government officials, and town dwellers in Palenque, in a corner of Mexico, who are prepared to assume this new awareness and behavior to solve the problem of deforestation locally and to avoid a global apocalypse. What they are asking for is fair: that, concurrent with taking on this problem, they should be given help to solve their other problems of poverty, unemployment, sickness, and despair. This requires local coordination, government efficiency, international financial assistance, scientific knowledge and technical skills, and fair trade in the exchange of goods between the North and the South. It is not an easy task. It implies the creation and enforcement of a political and ethical paradigm that for the first time in the history of humankind will have to involve all the Earth's inhabitants.

Understanding the world today calls for a creative, analytical, and effective science. And shaping global change requires hitherto unprecedented national and international negotiations, leading to the creation of new social and political charters for human coexistence.

Sustainability and human development should be made a top priority, as the United Nations is proposing today. They will form the basis for creating a set of goals that will provide access to a just global society, one based on cultural pluralism and on the current realities of the "wealth of nations" but with an additional mandate to consolidate the "wealth of humankind."

Appendix

Survey Methodology

In the survey 432 people were interviewed, including 24 women and 24 men of seven "communities" in two age cohorts (20–35 and 36–50). The wide age range of cohorts was due to the irregular age pyramid in the region, where the population over age 50 is extremely small because it is a largely immigrant population.

Within the age and gender groups a random sample was chosen in seven communities: two old settlements and two recent settlements in Marques de Comillas and five socioeconomic groups in Palenque.

Data from the fieldwork and in-depth interviews conducted during the initial period of fieldwork in 1990 allowed us to identify these communities. They were chosen on the basis of: (1) principal economic activity (and income level in Palenque), (2) type of settlement (rural or urban), and (3) duration of settlement (long established being in the 1970s, recent being between 1983 and 1988).

Accordingly, the following communities were chosen for the survey:

Community/Group	Characteristics
Pico de Oro	Old rain forest settlement
Reforma Agraria	Old rain forest settlement
Nuevo Chihuahua	Recent rain forest settlement
La Victoria	Recent rain forest settlement
La Unión	Old settlement near Palenque
El Lacandón	Old settlement near Palenque
Palenque bajos	Low-income group in Palenque
Palenque altos	High-income group in Palenque
Cattle ranchers	Cattle ranchers in Palenque

Notes

Chapter 1

1. In May 1990 the International Council of Scientific Unions (ICSU) held a meeting on "Science and Its Partners" to establish a dialogue between universities, private sector companies, and international science organization and governments on the challenges of global change.

2. T. Rosswall, personal communication, committee meeting on Human Dimensions of Global Change, April 1990, Paris.

3. Workshop on Global Change, sponsored by the Office for Interdisciplinary Earth Sciences, in which fifty natural, physical, and social scientists participated, 27 July–10 August, in Snowmass, Colorado.

4. The International Social Science Council, Standing Committee on Human Dimensions of Global Environmental Change. Jacobson and Price 1990. The committee is headed by Harold Jacobson and in its initial phase included Lourdes Arizpe, Daniel Bertaux, Ashish Bose, Takashi Fujii, Leszek Kosinski, Kurt Pawlik, Renat Perelet, Martin Price, and Robert Worcester.

5. "The Earth is one, but the world is not" is the phrase presented in the United Nations (UN) World Commission on Environment and Development report.

6. Daniel Bertaux discussed this point at one of the meetings of the ISSC UN and UNESCO Standing Committee on Human Dimensions of Global Change. We are grateful to the committee members for the debates, which provoked many of the reflections described in this book.

7. See also Gutman 1988; and Leff 1986.

8. International Council of Scientific Unions, the International Geosphere-Biosphere Program: A Study of Global Change, 1989, Paris.

9. We wish to acknowledge Benjamin Mayer's participation in the analysis in this section.

Chapter 2

1. Many of the features that José Bengoa (1981:163) attributes to the plantation system could equally be used to define the type of exploitation used by the rain forest timber firms: (1) foreign financing for the production and distribution of the product in question; (2) establishment of this type of firm in thinly

populated tropical zones; (3) use of labor recruited by coercive methods; (4) establishment of slave, semisalaried, or salaried relations according to the degree of development of the firm or region in question; (5) maximization of earnings through the exploitation of labor and the acquisition of vast tracts of land at practically no cost; (6) organization aimed at supplying a large-scale market; and (7) development based on monoculture (or monoproduction) to the detriment of the area.

2. Ironically, the expansion of agricultural lands in Mexico was due to López Mateos's Integral Agrarian Reform. He never imagined the consequences that this would have on the environment, as he himself admitted that he had refused to grant Maderera Maya S.A. permission to exploit the forest "so as not to go down in history as the man who destroyed the Lacandon rain forest."

Chapter 4

1. Taken from the definition of *perceive,* according to the *Diccionario de la Real Academía de la Lengua Española* 1984, 2:1041.

2. Ibid., 361.

3. For information on the survey, see appendix.

4. "War" is an important response because the survey was carried out at the time of the Persian Gulf War, a fact that should be taken into account.

5. "Other" included various responses, such as the lack of technology, natural disasters, the forest ban, and the forest's natural dangers.

6. Marijuana production has reached the Lacandona rain forest through drug traffickers, who have found the remote forest lots ideal for planting. It is said that it even has been planted in the Biosphere Reserve. Some farmers, although very few, are said to be involved, and certain ejidos in Marqués de Comillas are said to be centers of drug crops. This has led to forays and greater surveillance by the army. On its way from Palenque to the great markets of the north, marijuana has started to become easily available to the region's young people.

Chapter 5

1. Lacandons were not interviewed, since they do not live in the communities in which interviews were conducted. They live separately, in the depths of the forest, and have a very distinctive lifestyle.

Chapter 6

1. Arizpe 1978, 1985; Warman 1985.

2. Patrocinio González Garrido was governor of Chiapas from 1989 to 1993.

References

Abrams, Elliot R., and J. David Rue
   1988   The Causes and Consequences of Deforestation among the Prehistoric
          Maya. *Human Ecology* 16:377–95.
Alba, F., and J. Potter
   1986   Population and Development in Mexico since 1940: An Interpretation.
          Population and Development Review 12(1):47–75.
Appendini, K., and V. Salles, eds.
   1983   El crecimiento económico y el campesinado: Un análisis del ejido en
          dos décadas. In *El campesinado en México: Dos perspectivas de análisis.*
          129–254. Mexico City: El Colegio de México.
Argueta, A., and A. Embriz
   1990   Lacandona: Desastre o posibilidad. *La Jornada,* 19–22 August.
Aridjis, Homero
   1990   Montes Azules: Fin de la Lacandonia. *La Jornada,* 24–28 May.
Arizpe, L.
   1975   Problemas teóricos en el estudio de la migración de pequeños grupos:
          El caso de migrantes campesinos a la Ciudad de México. *Cahiers des
          Ameriques Latines* 12(2):201–22.
   1978   *Migración, etnicismo y cambio económico.* Mexico City: Colegio de
          México.
   1981   The Rural Exodus in Mexico and Mexican Migration to the United
          States. *International Migration Review* 15(4):626–50.
   1982   Relay Migration and the Survival of the Peasant Household. In *Why
          People Move,* ed. J. Balan. 19–31. Paris: UNESCO.
   1985   *Campesinado y migración.* Mexico City: (SEP).
   1989   On the Social and Cultural Sustainability of World Development. In
          One World or Several? ed. L. Emmerij, 207–19. Paris: OECD Develop-
          ment Center.
   1990   *La mujer en el desarrollo de México y de América Latina.* Mexico City:
          UNAM/Juan Pablos.
   1991   The Global Cube: Social Models in a Global Context. *International So-
          cial Science Journal* 130:599–608.
Arizpe, L., R. Costanza, and W. Lutz
   1991   Primary Factors Affecting Natural Resource Use. Document prepared
          for the International Council of Scientific Unions, ASCEND 21.

Asian Development Bank
  1990   Population Pressure and Natural Resource Management: Key Issues
         and Possible Actions. Paper no. 6.
Ayala, J., and J. Blanco
  1985   El nuevo estado y la expansión de las manufacturas. México, 1877–
         1930. In *Desarrollo y crisis de la economía mexicana*, ed. J. Ayala and J.
         Blanco, 14–44. Mexico City: Fondo de Cultura Económica.
Ayres, Robert
  1991   Production of Goods and Services: Towards a Sustainable Future. Talk
         presented at the symposium on "Humankind in Global Change: In-
         dicators and Prospects," organized by the American Association for
         the Advancement of Science Congress, February, Washington, D.C.
Bengoa, José
  1981   Plantaciones y agroexportación: Un modelo teórico. In *Desarrollo
         agrario y la América Latina*, ed. José Bengoa, 162–81. Mexico City: Fondo
         de Cultura Económica.
Blaikie, M., and B. Brookfield
  1987   *Land Degradation and Society*. London: Methuen.
Blaxter, Kenneth F. R. S.
  1986   *People, Food and Resources*. Cambridge: Cambridge University Press.
Boongarts, J., W. Mauldin, and J. Phillips
  1990   The Demographic Impact of Family Planning Programs. Population
         Council Research Division Working Paper no. 17.
Boserup, Esther
  1965   *The Conditions of Agricultural Growth: The Economics of Agrarian Change
         under Population Pressure*. Chicago: Aldine.
Bringley, Thomas
  1961   International Migration and Economic Development. Paris: UNESCO.
Brown, Harrison
  1954   *The Challenge of Man's Future*. New York: Viking Press.
Brown, Lester
  1990   La Ilusión de progreso. In *El Mundo: Medio Ambiente 1990*, ed. Lester
         Brown, 1–24. Mexico City: Fundación Universo Veintiuno.
Bruntland Commission
  1987   *Our Common Future*. New York: Oxford University Press.
Burton, I., and P. Timmerman
  1989   Human Dimensions of Global Change: A Review of Responsibilities
         and Opportunities. *International Social Science Journal* 121 (August):
         302.
Caldwell, John
  1984   Desertification: Demographic Evidence, 1973–83. Occasional Paper
         no. 37. Canberra: Australian National University.
Comisión Económica para América Latina (CEPAL)
  1991   *El desarrollo sustentable: Transformación productiva, equidad y medio am-
         biente*, 146. Santiago, Chile: CEPAL.

Chambers, R.
   1988   *Sustainable Livelihoods, Environment and Development: Putting Poor Peo-
           ple First.* Brighton, U.K.: Institute of Development Studies, University
           of Sussex.
Clark, Collin
   1958   Population Growth and Living Standards. In *The Economics of Under-
           development,* ed. A. N. Agarwal and S. P. Singh, 32–53. London: Oxford
           University Press.
Clark, William
   1991   Talk presented at the Annual Meeting of the American Association for
           the Advancement of Science. Washington, D.C., February.
Costanza, R., ed.
   1991   *Ecological Economics: The Science and Management of Sustainability.* New
           York: Columbia University Press.
DAWN-ISSC-SSRC
   1992   *Recasting the Population-Environment Debate: A Proposal for a Research
           Program.* Document produced in a meeting with the same name,
           Cocoyoc, Morelos, Mexico January 28, February 1, 1992.
Demeny, Paul
   1988   Demography and the Limits of Growth. *Population and Development
           Review,* suppl. 14:213–44.
   1990   Population. In *The Earth Transformed by Human Action: Global and Re-
           gional Changes in the Biosphere over the Past 300 Years,* ed. B. L. Turner et
           al., 41–54. New York: Cambridge University Press and Clark
           University.
Derrida, Jacques
   1978   *Writing and Difference.* London: Routledge and Kegan Paul.
Development and Environment Commission for Latin America and the
Caribbean
   1990   *Nuestra propia agenda sobne desarrollo y medio ambiente,* 102. BID-FCE-
           PNUD, Mexico City: UN.
De Vos, Jan
   1988   *Oro verde. La conquista de la Selva Lacandona por los madereros tabas-
           queños, 1822–1949,* 330. Mexico City: Fondo de Cultura Económica,
           Instituto de Cultura de Tabasco.
Dogan, Mattei, and Robert Pahre
   1989   Campos híbridos en las ciencias sociales. *Revista Internacional de Cien-
           cias Sociales* 121 (September): 497–512.
Douglas, Ian
   1991   *Human Settlements.* Document prepared for the 1991 Global Change
           Institute Workshop on Global Land-Use/Cover Change, organized by
           the Office for Interdisciplinary Earth Studies, 28 July–10 August,
           Snowmass, Colo.
Durning, Alan
   1991   Asking How Much Is Enough. In *State of the World 1991. A Worldwatch*

Institute Report on Progress toward a Sustainable Development, ed. L. R. Brown et al., 153–69. Washington, D.C.

Eckholm, R.

1982   *Down to Earth: Environmental and Human Needs.* New York: Norton.

Ehrlich, Paul, and Anne H. Ehrlich

1991   *The Population Explosion.* New York: Touchstone, Simon and Schuster.

Ehrlich, Paul, et al.

1989   Global Change and Carrying Capacity: Implications for Life on Earth. In *Global Change and Our Common Future: Papers from a Forum,* ed. Ruth DeFries and Thomas Malone, 19–27. Washington, D.C.: National Academy Press.

Food and Agriculture Organization (FAO)

1990   *Vital World Statistics.* Rome: FAO.

Favre, Henri

1984   *Cambio y continuidad entre los mayas de México. Contribución al estudio de la situación colonial en América Latina.* Mexico City: Instituto Nacional Indigenista.

Gallopín, Gilberto C.

1989   *Global Impoverishment, Sustainable Development and the Environment.* Buenos Aires: Ecological Analysis Group.

García Barrios, Raúl, Luis García Barrios, and Elena Alvarez-Buylla

1991   *Lagunas. Deterioro ambiental y tecnológico en el campo semiproletarizado.* Mexico City: El Colegio de México, Science Technology and Development Program.

García de León, A.

1985   *Resistencia y utopía. Memorial de agravios y crónica de revueltas y profecías acaecidas en la provincia de Chiapas durante los últimos quinientos años de su historia.* 2 vols. Mexico City: Editores ERA.

Gilland, Bernard

1983   Considerations on World Population and Food Supply. *Population and Development Review* 9(2):203–11.

1986   On Resources and Economic Development. *Population and Development Review* 12(2):295–305.

Goeller, H. E., and Alvin M. Weinberg

1976   The Age of Substitutability: What Do We Do When the Mercury Runs Out? *Science* 191:683–689.

González-Ponciano, R.

1990   Frontera, ecología y soberanía nacional. La colonización de la franja fronteriza sur de Marqués de Comillas. *Anuario del Instituto Chiapaneco de Cultura.* Tuxtla Gutiérrez, Mexico: Instituto Chiapaneco de Cultura.

González Pacheco, C.

1983   *Capital extranjero en la selva de Chiapas, 1863–1982.* Mexico City: UNAM.

Gordon, J., and T. Suzuki

1991   *It's a Matter of Survival.* Cambridge: Harvard University Press.

Grant, P., and M. Tanton
   1981   Immigration and the American Conscience. In *Progress as if Survival Mattered,* ed. A. Nash. Washington, D.C.: Friends of the Earth.
Gutelman, Michel
   1980   *Capitalismo y reforma agraria en México.* Mexico City: Editores ERA.
Gutman, Pablo
   1988   *Desarrollo rural y medio ambiente en América Latina.* Buenos Aires: CEAL.
Halperin, Tulio
   1986   Historia contemporánea de América Latina. Madrid: Alianza Editorial.
Hardoy, E., and W. Satterthwatte
   1991   Environmental Problems of the Third World Cities: A Global Issue Ignored? *Public Administration and Development* 11(2):45–62.
Hirsh, Fred
   1976   *Social Limits to Growth.* Cambridge: Harvard University Press.
Hirshman, Albert
   1958   *The Strategy of Economic Development.* New Haven: Yale University Press.
Jacobson, Harold K.
   1990   Note on the Human Dimensions for Global Change. HDGEC Internal Document no. 2.
Jacobson, Harold K., and Martin F. Price
   1990   *A Framework for Research on the Human Dimensions of Global Environmental Change.* Human Dimensions of Global Environmental Change Programme. Paris: International Social Science Council (ISSC)–UNESCO.
Johnson, D. Gale, and Lee D. Ronald, eds.
   1987   *Population Growth and Economic Development: Issues and Evidence.* National Research Council Committee on Population. Madison: University of Wisconsin Press.
Kelley, Allen
   1986   Review of the National Research Council Report on Population Growth and Economic Development: Policy Questions. *Population and Development Review* 12(3):563–67.
Keyfitz, Nathan
   1991   From Malthus to Sustainable Growth. International Institute for Applied Systems Analysis. July, Laxenburg, Austria.
Leff, Enrique
   1990   Población y medio ambiente. Es urgente detener la degradación ambiental. In *DEMOS. Carta demográfica sobre México.* 3:25–26.
   1992   La interdisciplinariedad en las relaciones población-medio ambiente. Hacia un paradigma de demografía ambiental. Talk presented at the Population and Environment Seminar, organized by the Sociedad Mexicana de Demografía, the Population Council, and the Red de

Formación Ambiental para América Latina y el Caribe-PNUMA-SEDUE, 8–10 April, Tepoztlán, Morelos.

Leff, Enrique, ed.
1986   *Los problemas del conocimiento y la perspectiva ambiental del desarrollo.* Mexico City: Editorial Siglo XXI.

Little, P., and M. Horowitz, with A. Endre Nyerges, eds.
1987   *Lands at Risk in the Third World: Local-Level Perspectives.* Boulder: Westview Press.

Lobato, Rodolfo
1979   Qu'ixin qui'nal. La colonización tzeltal de la selva lacandona. Undergraduate thesis. Mexico City: Escuela Nacional de Antropología e Historia.

Lutz, W., and C. Prinz
1991   Scenarios for the World Population in the Next Century: Excessive Growth or Extreme Aging. Working paper no. WP–91–22. Laxenburg, Austria: International Institute for Applied Systems Analysis.

Maihold, Gunter, and Victor Urquidi, comps.
1990   *Diálogo con nuestro futuro común. Perspectivas latinoamericanas del Informe Brundtland.* Friedrich Ebert Foundation, Mexico. Venezuela: Editorial Nueva Sociedad.

Maler, Karl-Goran
1990   Sustainable Development. Report of the Bergen Meeting on Environment. Norwegian Science Council.

Mandel, Ernest
1978   *Tratado de economía marxista.* Vol. 2. Mexico City: Editorial ERA.

Manzanilla, Linda
1989   *La formación del estado en Babilonia.* Mexico City: UNAM.

Meave del Castillo, J.
1990   *Estructura y composición de la selva alta perennifolia de los alrededores de Bonampak.* Mexico City: INAH.

Menon, G. K.
1989   Opening Address. In *Global Change,* Report no. 71. Paris: ICSSU.

Miller, Roberta B.
1989   Human Dimensions of Global Environmental Change. In *Global Change and Our Common Future,* ed. J. DeFries and T. Malone. Washington, D.C.: National Academy Press.

Muench, Pablo
1982   Las regiones agrícolas de Chiapas. *Geografía Agrícola* 2:57–102.

Murdock, George
1967   *Ethnographic Atlas.* Pittsburgh: University of Pittsburgh Press.

Myers, N.
1987   *Not Far Afield: U.S. Interests and the Global Environment.* Washington, D.C.: World Resources Institute (WRI).

Myers, Robert
1981   Deforestation in the Tropics: Who Gains, Who Loses? In *Where Have All*

the Flowers Gone? Studies in Third World Societies. Williamsburg, VA: Department of Anthropology, College of William and Mary.

National Council of Scientific Unions
   1989   *The International Geosphere-Biosphere Programme: A Study of Global Change*. Paris: International Council of Scientific Unions (ICSU).

Nigh, Ronald
   1992   La agricultura orgánica y el nuevo movimiento campesino en México. *Antropológicas* 3:29–50.

Organization of Economic Cooperation and Development (OECD)
   1991   *The State of the Environment*. Paris: OECD.

Ovando, Francisco Javier
   1981   La legislación forestal vigente. *Revista México Agrario* 14(3):43.

Pedrero, Gloria
   1983   El proceso de acumulación originaria en el agro chiapaneco. Siglo XIX. In *Investigaciones recientes en el área maya*, vol. 3:20. San Cristóbal de las Casas, Mexico: Sociedad Mexicana de Antropología.

Power, J.
   1992   *Informe sobre las mujeres rurales que viven en condiciones de pobreza*. Rome: Fondo Internacional de Desarrollo Agrícola.

Price, Martin
   1989   Global Change: Defining the "Ill-Defined." *Environment* 31(8):18–20, 42–44.

Przeworski, Adam
   1987   Marxismo y elección racional. *Zona Abierta* 45 (October–December): 30–35.

Repetto, Robert
   1987   *Population, Resources, Environment: An Uncertain Future*. Washington, D.C.: Population Reference Bureau.

Repetto, Robert, and Thomas Holmes
   1983   The Role of Population in Resource Depletion in Developing Countries. *Population and Development Review* 9(4): 1145–1163.

Revelle, Roger
   1976   The Resources Available for Agriculture. *Scientific American*, September, 165–78.

Reyes Osorio, S., et al.
   1974   *Estructura agraria y desarrollo agrícola en México*. Mexico City: Fondo de Cultura Económica.

Rockwell, Richard C.
   1992   Global Environmental Outcomes: The Current Scientific Understanding. Talk presented at the Population and Environment Workshop, organized by DAWN-ISSC-SSRC, 28 January–1 February, Cocoyoc, Morelos.

Sage, Colin, and M. Redclift
   1991   Population and Income Change: Their Role as Driving Forces of Land-Use Change. Document prepared for the 1991 Global Change Institute

Workshop on Global Land-Use/Cover Change, organized by the Office for Interdisciplinary Earth Studies, 28 July–10 August, Snowmass, Colo.

Sánchez, Vicente, Margarita Castillejos, and Leonora Rojas
1989   *Población, recursos y medio ambiente en México.* Mexico City: Fundación Universo Veintiuno, A.C.

Schmink, Marianne
1992   Deforestation: The Socioeconomic Matrix. Talk presented at the Population and Environment Workshop, organized by DAWN-ISSC-SSRC, 28 January–1 February, Cocoyoc, Morelos.

Secretaría de Agricultura y Recursos Hidráulicos (SARH)
1991   *Inventario nacional forestal de gran visión.* Mexico City: SARH.

Sen, Gita, and Karen Grown
1988   *Desarrollo, crisis y enfoques alternativos. Perspectivas de la mujer en el tercer mundo.* Mexico City: Piell/El Colegio de México.

Simon, Julian
1990   *Population Matters: People, Resource, Environment and Immigration.* New Brunswick, N.J.: Transaction Publishers.

Sloan, J., et al.
1988   *People of the Tropical Rainforest.* Berkeley: University of California Press.

Tabah, Léon
1992   Population Growth of the Third World. Document prepared for the International Conference on the Challenges to Systems Analysis in the Nineties and Beyond, organized by the International Institute for Applied Systems Analysis (IIASA), 12–13 May, Laxenburg, Austria.

Thompson, R., and L. Poo
1985   *Cronología histórica de Chiapas.* San Cristóbal de Las Casas, Chiapas: Centro de Investigaciones Ecológicas del Sureste.

Toledo, Víctor Manuel
1991   Modernidad y ecología. La nueva crisis planetaria. April. Mexico City.

Torrey, Barbara, and Gretchen Kolsrud
1991   Talk presented at the Annual Meeting of the American Association for the Advancement of Science, February, Washington, D.C.

Trent, John
1990   Thoughts on Political Thought: An Introduction. *International Political Science Review* 11:5–23.

Turner, B. L. II, et al.
1990   *The Earth Transformed by Human Action: Global and Regional Changes in the Biosphere over the Past 300 Years.* New York: Cambridge University Press and Clark University.

United Nations
1974   *Human Settlements: The Environmental Challenge.* London: MacMillan.
1990   *Global Outlook 2000: An Economic, Social and Environmental Perspective.* New York: United Nations Publications.

United Nations Department of International Economic and Social Affairs
1989   *World Population Prospects: 1988.* New York: Oxford University Press.

United Nations Development Programme
   1990 *Human Development Report: 1990.* New York: Oxford University Press.
   1992 *Human Development Report: 1992.* New York: Oxford University Press.
United Nations Population Fund
   1991a *Population, Resources and the Environment: The Critical Challenges.* London: Banson.
   1991b *The State of the World Population: 1991.* New York: Oxford University Press.
Velasco, Ciro
   1985 El desarrollo industrial de México en la década de 1930–1940. Las bases del proceso de industrialización. In *Desarrollo y crisis de la economía mexicana,* 45–65. Mexico City: Fondo de Cultura Económica.
Warman, Arturo
   1985 *Problemas agrarios.* Mexico City: Instituto Nacional Indigenista (INI).
Whitmore, Thomas M., et al.
   1990 Long-Term Population Change. In *The Earth Transformed by Human Action: Global and Regional Changes in the Biosphere over the Past 300 Years,* ed. B. L. Turner II et al., 25–39. New York: Cambridge University Press and Clark University.
Whyte, Anne V. T.
   1985 Perception. In *Climate Impact Assessment,* ed. R. W. Kates, J. H. Ausubel, and M. Berbertian, 403–36. Toronto: SCOPE, John Wiley and Sons.
Williams, Michael
   1991 Forest and Tree Cover. Document prepared for the 1991 Global Change Institute Workshop on Global Land-Use/Cover Change, organized by the Office for Interdisciplinary Earth Studies, 28 July–10 August, Snowmass, Colo.
Wittgenstein, Ludwig
   1958 Philosophical Investigations. 2d ed. Oxford: Basil Blackwell.
World Bank
   1992 *World Development Report 1992: Development and the Environment.* New York: Oxford University Press.
World Resources Institute
   1990 *World Resources 1990–91: A Guide to the Global Environment.* New York: Oxford University Press.
   1992 *World Resources 1992–93: A Guide to the Global Environment—Toward Sustainable Development.* New York: Oxford University Press.
Worldwatch Institute
   1988 *State of the World 1988: A Worldwatch Institute Report on Progress toward a Sustainable Society.* New York.